MEDICAL
INTELLIGENCE
UNIT

BLOOD GROUP ANTIGEN-RELATED GLYCOEPITOPES:
BIOLOGICAL FUNCTIONS AND IMPLICATIONS IN IMMUNOGENESIS, CANCER AND AIDS PATHOGENESIS

Gennadi V. Glinsky, M.D., Ph.D.

Cancer Research Center
Columbia, Mo.

R.G. LANDES COMPANY
AUSTIN

MEDICAL INTELLIGENCE UNIT

BLOOD GROUP ANTIGEN-RELATED GLYCOEPITOPES:
Biological Functions and Implications in Immunogenesis, Cancer and AIDS Pathogenesis

R.G. LANDES COMPANY
Austin / Georgetown

CRC Press is the exclusive worldwide distributor of publications of the Medical Intelligence Unit.
CRC Press, 2000 Corporate Blvd., NW, Boca Raton, FL 33431. Phone: 407/994-0555.

Submitted: September 1992
Published: November 1992

Production Manager: Terry Nelson
Copy Editor: Constance Kerkaporta

Please address all inquiries to the Publisher:
R.G. Landes Company
909 Pine Street
Georgetown, TX 78626
or
P.O. Box 4858
Austin, TX 78765
Phone: 512/ 863 7762
FAX: 512/ 863 0081

ISBN 1-879702-44-4
CATALOG # LN0244

Dedication

To my daughter Victoria and wife Anna with love and hope.

Acknowledgement

I would like to thank Drs. R.W. Zumwalt and C.W. Gehrke for their helpful advice and discussion; Mrs. Leanne McKnight for typing this manuscript; and my colleagues from the Academy of Science of Ukraine for many years of joint experimental work.

CONTENTS

CHAPTER 1

INTRODUCTION

The blood-group ABH determinants are major allogenic antigens in erythrocytes and many other human tissues. Historically, the term "blood-group antigens" (BGA) has been applied to the ABH and related antigens. However, they constitute major allogenic antigens of most epithelial cell types, and are also found in primary sensory neurons.[1,2] These antigens appear earlier in ectodermal or endodermal tissue than in mesenchymal hematopoietic tissues and cells.[2] Therefore, the term "histo-blood group" has recently been proposed for these allogenic carbohydrate antigens that generally represent peripheral parts of the oligosaccharide chains of glycoconjugates.[3] The structural aspects of the ABH antigens, relationships between their expression, development, differentiation and maturation, as well as malignant transformation, have been discussed.[3] It is well known that the expression of BGA-related glycodeterminants is greatly altered in many human cancers.[4-7] The first evidence of the occurrence of changes in BGA-glycoantigen expression in cancer was the indication more than 60 years ago of the incompatible expression of A antigen.[8-10] In the 1970s the appearance of new BG-A and BG-B antigens incompatible with the erythrocyte blood group of patients with carcinoma of the stomach and colon was observed by Hakkinen[11] and by Denk et al.[12] The importance of the problem of multiplicity of immunochemical, e.g., linked to the glycolipid or glycoprotein carrier, and biological, e.g., active or latent, forms of BGA-related glycodeterminants arises from the studies of the expression of BGA-related antigens in carcinomas of the gastrointestinal tract. Both the persistence,[12-17] loss or diminution,[18-22] of BGA-like activity in many gastrointestinal neoplasms has been described, and a direct relationship of the loss of BGA-reactivity to the anaplasia of the tumor as well as to its tendency to metastasize has been reported.[20-22] These discrepancies are due to the differences in methodology used and capability to detect water- and alcohol-soluble BGA, that are bound to glycoprotein and glycolipid carriers, correspondingly. Therefore, the different immunochemical isoforms of BGA display a remarkable difference in association with human cancer that may have a biological consequence determining the malignant behavior of tumor cells.

The first description of antigenic similarities between human malignant tumors and fetal tissues[23-25] has been followed by numerous reports confirming this finding in a variety of human neoplasms as well as in spontaneous and experimental animal tumors.[26-28] The carcinoembryonic antigen (CEA) and the α-fetoprotein were studied most intensively.[29,30] The blood group antigens have been considered as carcinofetal antigens in carcinomas of the distal colon[16] and a number of human cancer markers such as CEA contain the carbohydrate determinants of BGA. Stillman and Zamcheck[31] first discussed the possible relationship between the CEA and BGA. Immunochemical studies established a link between BGA and CEA[32] and a terminal N-acetylglucosamine has been considered as the key immunoreactive structure of CEA.[33] It has been suggested that the BGA-like carbohydrate site on the CEA molecule responsible for the immunologic cross-reactivity between CEA and BGA.[34-37] Furthermore, biochemical evidence indicates that oligosaccharide side chains of the CEA molecules contain the Lewis X trisaccharide,[38-40] LeY and occasionally a dimeric form of LeX antigens.[39,40]

Cancer cells frequently express carbohydrate embryonic antigens since they undergo a retrodifferentiation process during the course of malignant transformation.[41] SSEA-1 (stage-specific embryonic antigen-1) is one of the embryonic antigens which is specificaly expressed on the murine preimplantation embryos at the morula stage.[42] The SSEA-1 antigen is a carbohydrate antigen carrying LeX-hapten and i-antigenic structures.[43,44] Since SSEA-1 was first described as an embryonic antigen and subsequently found in various human cancers, the antigen has been regarded as an oncofetal or oncodevelopmental antigen.[45-48] It was shown that SSEA-1 antigen and its modified forms, including the sialylated forms (sialylated LeX or sialylated LeX-i) and fucosylated forms (LeY and poly LeX), are frequently accumulated in human cancer of various origins.[49-56] A large quantity of several fucose-containing glycolipids was found to be accumulated in various types of human adenocarcinomas, some of them were chemically and immunochemically identified as LeA, LeB, and LeX antigens.[57,58] In addition, other fucolipid analogs, such as LeY, sialyl LeA, disialyl LeA, and sialyl LeX, have been characterized as the tumor-associated glycodeterminants.[5] Associated with human cancer reduction of A or B determinants were first discovered by Oh-Uti in 1949[59] and subsequently confirmed by many others.[5] The pathophysiological significance of the deletion or reduction of A or B determinants, particularly related to the malignant poten-

tial of cancer cells, has been discussed.[7,60]

Finally, after introduction of monoclonal antibody (MAb) approach in cancer immunoglycobiology, it has been found that a number of antibodies selected on the basis of preferential reactivity with tumor cells over normal cells have been identified as being directed to the BGA-related glycoepitopes: LeX, LeA, LeB or their analogs,[5] and T-antigen.[61]

However, until recently the cell biological significance, if any, of the aberrant expression of BGA-related glycodeterminants on cancer cells was unclear, mainly because the physiological function, if any, of those glycoepitopes was unknown. Recent experimental evidence, that came from embryology, vascular cell biology, immunobiology and cancer cell biology, strongly indicates that the main biological role of the BGA-related glycodeterminants, in addition to their major contribution to the maintenance of the immunological homeostasis of the body, is the function as a key initial structural signal in cell-cell type specific recognition and adhesion.[62-65] It may well be that this dual function of BGA-related glycodetemrinants is glycoepitope density-dependent: Cells carrying a high density of BGA-related glycodeterminants will be eliminated, and cells carrying a low or medium level of BGA-related glycoepitopes will be the subject of clonal expansion and/or development of cell-cell contact. In this context, we will focus on the functional and pathophysiological aspects of blood-group antigen (BGA)-related glycoepitopes, mainly with regard to their key roles in cell-cell recognition, association, and adhesion, as well as on homotypic and heterotypic cell aggregation in relation to immunogenesis, cancer, and AIDS pathogenesis.

It has been postulated that in a broad range of histogenetically different tissues and organs, BGA-related glycoepitopes are expressed on the cell surface at definite stages of cell differentiation. These glycoepitopes are expressed during embryogenesis, organogenesis, tissue repair, regeneration, remodeling and maturation when "sorting-out" of one homotypic cell population from a heterotypic assemblage of cells occurs.[62,63] In this event, the BGA-related glycoepitopes, if being expressed on the cell surface, play roles of key structural determinants in cell-cell recognition, association and aggregation. This mechanism will be discussed in relation to immunogenesis with regard to antigen presentation, self-non-self discrimination, and positive and negative selection during thymic education. It is postulated that the appearance of BGA-related glycoepitopes on the cell membrane is a consequence of the association of major histocompatibility complex antigens (MHC) and peptides, with the subsequent elimination of cells carrying a high density of BGA-related glycoepitopes on their surface.[62-65] After human immunodefficiency virus (HIV) glycoproteins are glycosylated by host cell glycosyltransferases, the virus may use the BGA-related glycodeterminants as ligands and/or receptors for expansion to a spectrum of target cells during AIDS development and generalization of the infection throughout the body.[62,64]

We will review the experimental evidence that supports the concept that HIV uses an alternative to the gp120/CD4 ligand/receptor system, and that the alternative mechanism is probably carbohydrate-mediated in nature. It has been speculated that during AIDS development, HIV could use a "receptor-driving" mechanism, e.g., change of the predominant protein-protein type of ligand-receptor interaction at the first stage of infection to the carbohydrate-protein, or carbohydrate-carbohydrate types of virus ligand-target cell receptor interactions at the later stage of infection.[62,64] The role in these processes of extracellular, in particular serum biomacromolecules, carrying and/or specifically recognizing BGA-related glycoepitopes, will be considered, and the possibility of applying the BGA-related glycoepitopes themselves as tools for anticarbohydrate vaccine therapy will be suggested.

In cancer, the expression of BGA-related glycoepitopes has been considered a key mechanism of phenotypic divergence of tumor cells, immunoselection, tumor progression, and metastasis.[62,65] The cell-biological and molecular mechanism of site specificity of cancer metastasis remains a mystery. The traditional lock-and-key model predicts that cancer cells may have a specific membrane-associated structure(s) which determines specificity of cancer cell-endothelium recognition and site(s) of metastatic deposition. Here, we will summarize the evidence for an alternative concept. According to this model, the site specificity of cancer metastasis is determined by leukocyte-endothelial cell recognition and adhesion. The homotypic and heterotypic cancer cell aggregation in blood vessels and subsequent formation of "multicellular metastatic units" will be considered as prerequisite to cancer metastasis.

This "multicellular metastatic unit" consists of tumor cells, leukocytes, and platelets, where site specificity of endothelium recognition, adhesion and stable attachment will be determined by a subset(s) of leukocytes involved in the formation of the "multicellular metastatic unit". This particular subset of leukocytes may serve as "carrier cells" targeting the metastatic deposit(s) to the specific site(s) of secondary tumor foci formation. The prospect of anti-cell adhesion therapy of cancer metastasis [62,63,65-68] will be discussed with emphasis on the inhibition of carbohydrate-mediated homo- and heterotypic cancer cell aggregation.

Cell-cell type specific recognition, association and adhesion are the key critically important events in the development (embryo-, morpho- and organogenesis), immunological response, inflammation, and pathogenesis of human diseases such as cancer, atherosclerosis, immunodeficiency syndrome, including AIDS. When I first discussed this concept during the 5th and 6th International Conferences of International Academy of Tumor Markers Oncology (IATMO) in Stockholm (September, 1988), and Tokyo (May, 1989), and in Freiburg (September, 1989) during XVIIth meeting of International Society of Oncodevelopmental Biology and Medicine (ISOBM) it was unclear how soon these theoretical considerations would have some practical implications.

During subsequent series of scientific

seminars in the USA, Japan, Austria and Germany, I have concluded that the dramatic progress in our understanding of the role of specific carbohydrate determinants in early step multicomponent and multistep cellular adhesion mechanisms indicates that we are entering a new era of the pathogenetic cell type specific treatment of human diseases.

This treatment strategy will be based on a specific carbohydrate- associated cell adhesion inhibitory mulecules, and will become a reality by the end of 1992. Cytel, Inc. is reported to be enthusiastic in its development of a carbohydrate that mimics the sialylLeX structure to block the selectin receptor sites. One of the compounds is slated for clinical trials in late 1992 for acute antiinflammatory disease, and a carbohydrate based leukocyte-endothelial cell adhesion blocker is expected to be a subject of study as an antiinflammatory drug in 1993 to treat adult respiratory distress syndrome. Glycomed, Inc. and Eli Lilly have entered into collaboration, that is due to expire in January 1993, and is directed toward novel oligosaccharide compounds which may be useful in the prevention and treatment of vascular restenosis and atherosclerosis. In 1992 an agreement was signed between the Alberta Research Council and Glycomed for development of carbohydrate based treatment for inflammatory and immune diseases.

References

1. Hakomori, S.I. (1981) Blood group ABH and Ia antigens of human erythrocytes: Chemistry, polymorphism and their developmental change, Semin Hematol. 18:39-62.
2. Oriol, R., LePendu, J., Mollicone, R. (1986) Genetics of ABO, H, Lewis, X and related antigens. Vox Sang. 51:161-171.
3. Clausen, H., Hakomori, S.I. (1989). ABH and related histo-blood group antigens; Immunochemical differences in carrier isotypes and their distribution. Vox Sang. 56, 1-20.
4. Hakomori, S. (1984). Philip Levine Award Lecture: Blood group glycolipid antigens and their modification as human cancer antigens. Am. J. Clin. Pathol. 82:635-648.
5. Hakomori, S. (1989). Aberrant glycosylation in tumors and tumor-associated carbohydrate antigens. Adv. Cancer Res. 52:257-331.
6. Kuhns, W.J., Primus, F.J. (1985). Prog. Clin. Biochem. Med. 2:49-95.
7. Feizi, T. (1985). Carbohydrate antigens in human cancer. Cancer Surv. 4:245-269.
8. Witebsky, E. (1929). Z. Immunitaetsforsch. Exp. Ther. 62:35-73.
9. Hirszfeld, L., Halber, W., Laskowsky, J. (1929). Z. Immunitaetsforsch. Exp. Ther. 64:81-113.
10. Witebsky, E. (1930). Zur serologischen spezifitat des carcinomgewebes. Klin. Wochenschr. 9:58-63.
11. Hakkinen, I. (1970). A-like blood group antigen in gastric cancer cells of patients in blood groups O or B. J. Natl. Cancer Inst., 44:1183-1193.
12. Denk, H., Tappeiner, G., Davidovits, A., Eckerstorfer, R., Holzner, J.H. (1974a) Carcinoembryonic antigen and blood group substances in carcinomas of the stomach and colon. J. Natl. Cancer Inst., 53:933-942.
13. Glynn, L.E., Holborow, E.J., Johnson, G.D. (1957). The distribution of blood-group substances in human gastric and duodenal mucosa. Lancet, 2:1083-1088.
14. Glynn, L.E., Holborow, E.J. (1959). Distribution of blood group substances in human tissues. Br. Med. Bull., 15:150-153.
15. Eklund, A.E., Gullbring, B., Lagerlof, B. (1963). Blood group specific substances in human gastric carcinoma. A study using fluorescent antibody technique. Acta. Pathol. Microbiol. Scand., 59:447-455.
16. Denk, H., Tappeiner, G., Holzner, J.H. (1974b). Blood group substances (BG) as carcinofetal antigens in carcinomas of the distant colon. Europ. J. Cancer, 10:487-490.
17. Denk, H., Tappeiner, G., Holzner, J.H. (1974c). Independent behaviour of blood group A- and B-like activities in gastric carcinomata of blood group AB individuals. Nature, 248:428-430.
18. Cowan, W.K. (1962). Blood group antigen on human gastrointestinal carcinoma cells. Br. J. Cancer, 16:535-540.
19. Sheahan, D.G., Horowitz, S.A., Zamcheck, N. (1971). Deletion of epithelial ABH isoantigens in primary gastric neoplasms and metastatic cancer. Am. J. Dig. Dis., 16:961-969.
20. Davidsohn, I., Kovarik, S., Lee, C.L. (1966). A,B,O substances in gastrointestinal carcinoma. Arch. Pathol., 81:381-390.
21. Davidsohn, I., Ni, L.Y., Stejskal, R. (1971). Tissue isoantigens A,B and H in carcinoma of the stomach. Arch. Pathol., 92:456-464.
22. Davisohn, I. (1972). Early immunologic diagnosis and prognosis of carcinoma. Am. J. Clin. Pathol., 57:715-730.
23. Hirszfeld, L. (1929). Untersuchungen uber die serologischen Eigenschaften der Gewebe: uber serologische Eigenschaften der Neubildungen. Z. Immun. Forsch. Exp. Ther., 64:81.
24. Hirszfeld, L., Halber, W., Floksstrumpf, M., Kolodzieski, J. (1930). Uber Krebsantikorper bei Krebskranken. Klin. Wschr. 9:342-345.
25. Hirszfeld, L., Halber, W., Rosenblat, J. (1932). Untersuchungen uber verwandt-schaftsreaktionen

zwischen embryonalund krebsgewebe. Z. Immunforsch., 75:209-216.

26. Gold, P. (1971). Embryonic origin of human tumor-specific antigens. Prog. Exp. Tumor Res., 14:43-58.

27. Laurence, D.J.R., Neville Munro, A. (1972). Foetal antigens and their role in the diagnosis and clinical management of human neoplasms: A review. Brit. J. Cancer, 26:335-355.

28. Stonehill, E.H., Bendich, A. (1970). Retrogenic expression: the reappearance of embryonal antigens in cancer cells. Nature, 228:370-372.

29. Gold, P., Freedman, S.O. (1965). Specific carcinoembryonic antigens of the human digestive system. J. Exp. Med., 122:467-481.

30. Abelev, G.I. (1968). Production of embryonal serum α- globulin by hepatomas. Review of experimental and clinical data. Cancer Res., 28:1344-1350.

31. Stillman, A., Zamcheck, N. (1970). Recent advances in immunologic diagnosis of digestive tract cancer. Am. J. Dig. Dis., 15:1003-1018.

32. Simmons, D.A., Perlmann, P. (1973). Carcino-embryonic antigen and blood group substances. Cancer Res., 33:313-322.

33. Banjo, C., Gold, P., Freedman, S.O., Krupey, J. (1972). Immunologically active heterosaccharides of carcinoembryonic antigen of human digestive system. Nature, New Biol., 238:183-185.

34. Gold, J.M., Freedman, S.O., Gold, P. (1972). Human anti-CEA antibodies detected by radio-immunoelectrophoresis. Nature, New Biol., 239:60-62.

35. Gold, J.M., Gold, P. (1973). The blood group A-like site on the carcinoembryonic antigen. Cancer Res., 33:2821-2824.

36. Gold, J.M., Banjo, C., Freedman, S.O., Gold, P. (1973). Immunochemical studies of the intramolecular heterogeneity of the carcinoembryonic antigen (CEA) of the human digestive system. J. Immunol., 111:1872-1879.

37. Turner, M.D., Olivares, T.A., Harwell, L., Kleinman, M.S. (1972). Further purification of perchlorate-soluble antigen from human colonic carcinomata. J. Immunol., 108:1328-1339.

38. Chandrasekaran, E.V., Davila, M., Nixon, D.W., Goldfarb, M., Mendicino, J. (1983). Isolation and structures of the oligosaccharide units of carcinoembryonic antigen. J. Biol. Chem., 258:7213-7222.

39. Nichols, E.J., Kannagi, R., Hakomori, S., Krantz, M.J., Fuks, A. (1985). Carbohydrate determinants associated with carcinoembryonic antigen (CEA). J. Immunol., 135:1911-1913.

40. Yamashita, K., Totani, K., Kuroki, M., Matsuoka, Y., Ueda, I., Kobata, A. (1987). Structural studies of the carbohydrate moieties of carcinoembryonic antigens. Cancer Res., 47:3451-3459.

41. Miyake, M., Zenita, K., Tanaka, O., Okada, Y., Kannagi, R. (1988). Stage-specific expression of SSEA-1-related antigens in the developing lung of human embryos and its relation to the distribution of these antigens in lung cancers. Cancer Res. 48:7150-7158.

42. Solter, D., Knowles, B.B. (1978). Monoclonal antibody defining a stage-specific mouse embryonic antigen (SSEA-1). Proc. Natl. Acad. Sci. USA 75:5565-5569.

43. Gooi, H.C., Feizi, T., Kapadia, A., Knowles, B.B., Solter, D., Evans, M.J. (1981) Stage-specific embryonic antigen involves alpha 1 goes to 3 fucosylated type 2 blood group chains. Nature. 292:156-158.

44. Kannagi, R., Nudelman, E., Levery, S.B., Hakomori, S. (1982) A series of human erythrocyte glycosphingolipids reacting to the monoclonal antibody directed to a developmentally regulated antigen SSEA-1. J. Biol. Chem. 257:14865-14874.

45. Andrews, P.W., Goodfellow, P.N., Shevinsky, L.H., Bronson, D.L., Knowles, B.B. (1982) Cell-surface antigens of a clonal human embryonal carcinoma cell line: morphological and antigenic differentiation in culture. Int. J. Cancer. 29:523-531.

46. Combs, S.G., Marder, R.J., Minna, J.D., Mulshine, J.L., Polovina, M.R., Rosen, S.T. (1984) J. Histochem. Cytochem. 32:982-988.

47. Fox, N., Damjanov, I., Martinez-Hernandez, A., Knowles, B.B., Solter, D. (1981) Immunohistochemical localization of the early embryonic antigen (SSEA-1) in postimplantation mouse embryos and fetal and adult tissues. Dev. Biol. 83:391-398.

48. Fox, N., Damjanov, I., Knowles, B.B., Solter, D. (1983) Immunohistochemical localization of the mouse stage-specific embryonic antigen 1 in human tissues and tumors. Cancer Res. 43:669-678.

49. Abe, K., J.M. McKibbin and S.I. Hakomori. (1983) The monoclonal antibody directed to difucosylated type 2 chain (Fuc alpha 1 leads to 2Gal beta 1 leads to 4 [Fuc alpha 1 leads to 3 GlcNAc; Y Determinant). J. Biol. Chem. 258:11793-11797.

50. Fukushi, Y., Kannagi, R., Hakomori, S., Shepard, T., Kulander, B.G., Singer, J.W. (1985) Location and distribution of difucoganglioside in normal and tumor tissues defined by its monoclonal antibody FH6. Cancer Res. 45:3711-3717.

51. Hakomori, S., Nudelman, E., Levery, S.B., Kannagi, R. (1984) Novel fucolipids accumulating in human adenocarcinoma. I. Glycolipids with di- or trifucosylated type 2 chain. J. Biol- Chem 259(7):4672-4680

52. Kannagi, R., Fukushi, Y., Tachikawa, T., Noda, A., Shin, S., Shigeta, K., Hiraiwa, N., Fukuda, Y., Inamoto, T., Hakomori, S., Imura, H. (1986) Quantitative and qualitative characterization of human cancer-associated serum glycoprotein antigens expressing fucosyl or sialyl-fucosyl type 2 chain polylactosamine. Cancer Res. 46:2619-2626.

53. Shi, Z.R., McIntyre, L.J., Knowles, B.B., Solter, D., Kim, Y.S. (1884) Expression of a carbohydrate differentiation antigen, stage-specific embryonic antigen 1, in human colonic adenocarcinoma. Cancer Res. 44:1142-1147.

54. Spitalnik, S.L., Spitalnik, P.F., Dubois, C., Mulshine, J., Magnani, J.L. Cuttitta, F., Civin, C.I., Minna, J.D., Ginsburg, V. (1986). Glycolipid antigen expression in human lung cancer. Cancer Res. 46:4751-

4755.

55. Cuttita, F., Rosen, S., Gazdar, A.F., Minna, J.D. (1981) Proc. Natl. Acad. Sci. USA 78:4591-4595.

56. Huang, L.C., Brockhaus, M., Magnani, J.L., Cuttitta, S.R., Ginsburg, V. (1983) Many monoclonal antibodies with an apparent specificity for certain lung cancers are directed against a sugar sequence found in lacto-N-fucopentaose III. Arch. Biochem. Biophys. 220:318.

57. Hakomori, S., Andrews, H. (1970). Sphingoglycolipids with LeB acitivity and the copresence of LeA-, LeB- glycolipids in human tumor tissue. Biochim. Biophys. Acta. 202:225-228.

58. Yang, H.-J., Hakomori, S. (1971). A sphingolipid having a novel type of ceramide and lacto-N-fucopentaose III., J. Biol. Chem., 246:1192-1200.

59. Oh-Uti, K. (1949). Polysaccharides and a glycidamin in the tissue of gastric cancer. Tohoku J. Exp. Med. 51:297-304.

60. Orntoft, T.F., Wolf, H., Clausen, H., Hakomori, S., Dabelsteen, E. (1987). Blood group ABO and Lewis antigens in fetal and normal adult bladder urothelium: Immunohistochemical study of type 1 chain strucutre., J. Urol. 138:171-176.

61. Stein, R., Chen, S., Grossman, W., Goldenberg, D.M. (1989). Human lung carcinoma monoclonal antibody specitic for the Thomsen-Friedenreich Antigen. Cancer Res. 49:32-37.

62. Glinsky, G.V. (1992a). The blood group antigens (BGA)-related glycoepitope. A key structural determinants in immunogenesis and cancer pathogenesis. Critical Reviews in Oncology/Hematology, 12:151-166.

63. Glinsky, G.V. (1992b). Glycoamines: Structural-functional characterization of a new class of human tumor markers. In: Serological Cancer Markers. Editor: S. Sell. The Humana Press, Totowa, N.J., Chapter 11, pp. 233-260.

64. Glinsky, G.V. (1992c). The blood group antigen related glycoepitopes: Key structural determinants in immunogenesis and AIDS pathogenesis. Medical Hypotheses (in press).

65. Glinsky, G.V. (1992d). The site specificity of cancer metastasis: Is it determined by leukocyte-endothelial cell recognition and adhesion? (in preparation).

66. Glinsky, G.V. (1992e). Glycodeterminants of melanoma cell adhesion. A model for antimetastatic drugs design. Critical Reviews in Oncology/Hematology (submitted).

67. Hakomori, S. (1991). Possible functions of tumor-associated carbohydrate antigens. Current Opinion in Immunology., 3:646-653.

68. Hakomori, S. (1992). Possible new directions in cancer therapy based on aberrant expression of glycosphingolipids: Antiadhesion and ortho-signaling therapy. Cancer Cells (in press).

CELL ADHESION AS A MULTICOMPONENT AND MULTISTEP "CASCADE" PROCESS

Interacting cells coexpress on their surface the multiple families of different molecules involved in cell-cell type specific recognition, association and adhesion. The function of multiple cell adhesion molecules on a single cell during specific cell-cell adhesion has been explored in several experimental systems.[1-4] It has been shown, for example, that the extension of neurites by peripheral motor neurones on the surface of myotubes involves multiple mechanisms for growth and guidance acting in concert during development.[2] This process involves neuronal extracellular matrix receptors (particularly, laminin), as well as Ca^{2+}-dependent (neural Ca^{2+}-dependent cell adhesion molecules) and Ca^{2+}-independent (neural cell adhesion molecules) cell adhesion molecules on neurite surfaces. The neuronal receptors for extracellular matrix proteins and different cell adhesion molecules, used by neuronal growth cones for interaction with myotubes, each provide a distinct but cooperative mechanism for regulation of growth cone motility.[2] One constituent of such a multicomponent cell adhesion system is the cell membrane carbohydrates. Depending on structure of carbohydrates they may act either as inhibitor or activator of cell-cell recognition and adhesion. The function of the neural cell adhesion molecule (NCAM), one of the most abundant adhesion molecules, can influence a wide variety of diverse cellular events: junctional communication, the axon association with pathways and targets and signals that alter levels of neurotransmitter enzymes.[1]

These pleiotropic effects reflect the ability of NCAM to regulate membrane-membrane contact of interacting cells required for initiation of specific interactions between other molecules. Such regulation can occur through changes in either NCAM expression or the content of polysialic acid (PSA) on NCAM molecules.[1] When NCAM with a low PSA content is expressed, adhesion is increased, the probability of junction formation from interacting subunits of other adhesion molecules is enhanced by increasing the extent or duration of membrane-membrane contact and contact-dependent events are triggered. The large excluded volume of NCAM PSA may inhibit cell-cell interaction through hindrance of overall membrane apposition,[1] and the initiation of cellular interactions via specific ligand-receptor pairs may occur as a result of reduction in NCAM PSA content. The probability of receptor-ligand interactions is enhanced by reducing the excluded volume of carbohydrate between membranes, which otherwise impedes close cell-cell contact. These steric alterations of the overall degree of close membrane-membrane contact could allow differential regulation of the function of several ligand-receptor systems according to their properties, such as size, mobility and quantity on the cell surface, etc.[1]

The process of cell-cell recognition, association and aggregation is not only multicomponent, but consists of multiple steps and one of the models of such a multistep process was proposed by S. Hakomori and coworkers.[5] as an explanation some of the development-related phenomena. The initial step in the process is specific recognition between two cells in which multivalent homo- and heterotypic carbohydrate-carbohydrate and/or carbohydrate-lectin interactions may play a major role.[6] Initial cell recognition through carbohydrate-carbohydrate or carbohydrate-protein (selectin) interaction is followed by nonspecific adhesion. This nonspecific adhesion is primarily mediated by Ca^{2+}-sensitive adhesion molecules or sugar-binding proteins, or by proteins of the immunoglobulin superfamily, or by pericellular adhesive meshwork proteins consisting of fibronectin, laminin and their receptor systems (integrin). The third step of cell adhesion is the establishment of intercellular junctions, e.g., "gap junctions," in which a cell-cell communication channel is opened through a specific structural protein. The "sorting-out" behavior of homotypic cell populations plays a major role in morphogenesis and organogenesis and in this context cell adhesion is processed mainly through recognition between similar types of cells. Specific interactions between identical or similar carbohydrate chains which are highly expressed in the same types of cells may provide a better explanation for sorting than the interaction of relatively nonspecific adhesive molecules.[5] It is important to note that tumor cells retain, to a great extent, their ability to reveal initial stages of cell adhesion: at the same time cancer is characterized by profound disturbances in the subsequent stages of cell adhesion, with occurence of involvement of extracellular matrix proteins and followed by the formation of specific "gap junctions." We suggested[7-11] that the recog-

nition between identical types of cells, that provides the basis for sorting a single homotypic population from a heterotypic assemblage of cells, plays a major role in the metastatic spreading of tumor cells. This mechanism provides a basis for the formation of tumor cell multicellular aggregates in the circulatory channels of the tumor host and the subsequent arrest of these multicellular aggregates in blood vessels of target organs.

The carbohydrate determinants of the BGA-related glycoantigens may be the structural determinants participating in the homotypic aggregation of histogenetically different types of cells.[11] BGA-related glycoepitopes are directly involved in the homotypic (tumor cells, embryonal cells) and heterotypic (tumor cells-normal cells) formation of cellular aggregates, e.g., Lewis X antigens; H-antigens, polylactosamine sequences; and T- and Tn-antigens, which was demonstrated in different experimental systems.[12-17] Lymphoid cells (lymphocytes, granulocytes, monocytes) are Lewis family glycoepitope-positive and contain T-antigen (demasked after treatment with neuraminidase).[18-20] Carbohydrate ligands for selectins—major adhesion proteins of circulating cells, e.g., leukocytes, platelets, and endothelium—have recently been discovered: sialosyl LeX, sialosyl LeA and LeX have been identified as recognition structures for selectins.[21-24] Neutrophils bear LeX both on glycolipids and at the termini of N- and O-linked oligosaccharides.[25,26] The selectin ELAM-1 is involved in the differential attachment to endothelium and subsequent migration into the tissues of specific lymphocyte subsets (CD4+ memory T-cells).[27,28]

Recent experimental evidence has generally supported the concept that some of the BGA-related glycodeterminants, which were identified earlier as tumor-associated carbohydrate antigens (TACA), function as key adhesion molecules.[11,29,30] Recent studies have shown that cell adhesion, through carbohydrate-carbohydrate or carbohydrate-selectin interactions, occur at the early initial stages of the "cascade" multistep cell adhesion mechanism and this reaction is a prerequisite for subsequent steps of cell adhesion directed at integrin- or immunoglobulin-based adhesion.[31,32] Usually, cells co-express on their surface the multiple components involved in the "cascade" cell adhesion mechanism and thus this multistep adhesion reaction could be triggered by initial carbohydrate-carbohydrate or carbohydrate-selectin interactions. Such conclusions have been made on the basis of studies of B16 melanoma cell adhesion[30,31] and investigations of the leukocyte's role on a selectin.[32] Evidence has been presented that specific glycosphingolipid-glycosphingolipid interaction initiates cell-cell adhesion and may cooperate synergistically with other cell adhesion systems such as those involving integrins.[29-31,33]

A GENERAL MODEL OF CELLULAR ADHESION

Thus, specific membrane-associated carbohydrate determinants, particularly BGA-related glycoepitopes, function as a ligand and/or receptor in cell-cell recognition, association, aggregation and adhesion. Initial carbohydrate-mediated cell-cell recognition and interaction may trigger the subsequent steps of the "cascade" cell adhesion mechanism and involve the multiple family of cell adhesion molecules in the formation of the stable and specific cell-cell contact. It has been suggested that specific glycosphingolipid-glycosphinglolipid interactions may be the general mechanism of the initiation of cell-cell adhesion.[29-31,33] This model may be correct for some of the specific cell type, like melanoma cells, that contain an unusually large amount of gangliosides on the cell surface. However, in the most cases the specific cell-cell recognition has to be initiated by transmembrane glycoproteins with relatively large size extracellular carbohydrate-carrying and/or -binding domains.

On a molecular scale a model of the distances over which adhesion receptors would mediate cellular interactions has been constructed.[34] This model allowed prediction of how close two cell membrane, or a cell membrane and the extracellular matrix, would have to come to allow effective interactions. There are two classes of adhesion interactions that differ significantly in the distance between

the plasma membrane of the two closely apposed cell.[34] The first type of cell-cell adhesion contact is predicted to occur within an intermembrane distance roughly equal to 30 nm and the second class of intercellular adhesion interactions is predicted to occur within an intermembrane distance of 10-15 nm, which is in good agreement with electron-microscopic measurments of intermembrane distance of membrane apposition in interactions of mitogen-stimulated T cells with target cells.[35] For example, migrating cells do not possess focal contacts, which have a cell to substrate distance of 10-15 nm, but have close contacts with a cell to substrate distance of 30 nm,[36] in good agreement with the distance predicted for FLA-1/ICAM interaction and for integrin-matrix interaction.[34] Furthermore, the scale of glycocalyx has to be considered, which will oppose cellular interaction due to negative charge repulsion and the loss of entropy involved in compressing or interdigitating the glycocalyces of two contacting cells.[37] The sizes of two major components of the leukocyte glycocalyx, that bear most of the cell surface sialic acid,[38] is about 30 nm.[34] On the time scale it is more likely that the first type of cell-cell adhesion contact will occur at the early initial stage of cell-cell recognition and adhesion and this step more likely will be mediated by transmembrane glycoproteins with a relatively big extracellular domain. The concept of the crypticity of glycolipid antigens supports this suggestion.

According to the minimum energy conformation model, the axis of the carbohydrate is perpendicular to that of ceramide; glycolipids are inserted in the lipid bilayer through their ceramide moiety and their carbohydrates are laid on and fixed to the surface of the lipid bilayer, exposing the hydrophilic surface toward the outside of the membrane.[39,40] According to this model, crypticity of glycolipid antigens will be determined and controlled by several complex factors, such as membrane proteins and sialosyl glycoconjugates surrounding glycolipids, ceramid composition, etc.[41] Therefore, glycolipids are more likely to be involved on the late stages of cell-cell adhesion when cell-cell contact distance between the plasma membranes of the two closely apposed cells is 10 nm or less. One important question regards the regulatory mechanism responsible for sequential cooperative action of the different cellular adhesion systems during the cell-cell recognition, association and stable attachment.

Considering the molecular scale and time scale models, the molecular topography of primary adhesion molecules as well as their rapid down-regulation after initiation of cell-cell contact appear to be critically important. For example, neutrophil extravasation at sites of acute inflammation may be considered a two-step process.[42,43] First, neutrophils within the circulating blood must recognize and bind venous endothelial cells activated by a tissue-based inflammatory stimulus. The second step involves adhesion strengthening, ensuring the complete stop of the rolling neutrophils, followed by transmigration through the vascular wall. The leukocyte selectin (LECAM-1) and the vascular selectins (ELAM-1 and GMP-140) are primarily responsible for the initial recognition/binding interaction of neutrophils and endothelial cells[32,43-45] The β2 integrin adhesion pathway is primarily involved in the second step of this process and appears critically involved in the adhesion strengthening and transmigration stages of the extravasation process.[44,46-49]

The expression of LECAM-1 on neutrophils is extremely sensitive to neutrophil activation by cytokines, chemoattractants and other inflammatory agents:[50-52] Complete shedding of LECAM-1 occurs within five minutes of exposure of murine neutrophils to recombinant C5a at 37°C.[50] The same factors that cause LECAM-1 shedding also cause up-regulation of β2 integrin function.[47,48] After initial interaction of neutrophils with endothelial cells, largely selectin mediated, LECAM-1 shedding would occur, down-regulating a major portion of the selectin-mediated adhesion, to facilitate or trigger the integrin-mediated stable attachment and subsequent diapedesis.[42] Down-regulation of selectin function by LECAM-1 shedding may also serve as an autoregulatory mechanism limiting or blocking excessive neutrophil-endothelial cell interaction after severe inflammatory stimuli.

The critical importance of molecular to-

pography of adhesion molecules for initiation of cell-cell contact has been shown by Picker et al.[42] The neutrophil selectin LECAM-1 participates in the earliest interactions of neutrophils with endothelial cells through presentation of oligosaccharide ligands to the inducible vascular selectins (ELAM-1 and GMP-140). The enhanced function of LECAM-1 associated oligosaccharides (sialosyl LeX) reflects the striking concentration of LECAM-1 on neutrophil surface microvilli, the site of initial cellular contact.[42] One mechanism of the cooperative action of different family of adhesion molecules is the processes designated as mutual capping and/or mutual co-capping.[53,54]

The initial binding of a cell adhesion mulecule to its receptor under appropriate conditions may result in dramatic redistribution of integral membrane molecules within the fluid membranes of the two contacting cells.[55] The concentration of the ligand and receptor molecules should be sufficiently large and the intrinsic rate constant for the dissociation of the ligand-receptor bond should be sufficiently small to allow the formation of a small number of initial ligand-receptor bonds which can maintain localized cell-cell contact long enough for the diffusion of many more ligand and receptor molecules of the same specificity (mutual capping) or a variety of different classes of cell adhesion molecules that contact cells may simultaneously exhibit (mutual co-capping). These ligand and receptor molecules after diffusion into the contact region will form additional bonds clustered within an extended contact area. The function of mutual capping and co-capping is the increase in the concentration of corresponding cell adhesion molecules precisely within the morphologically defined region of cell-cell contact.[54]

This clustering phenomenon occurs because the formation of the initial cell-cell contact as a result of interaction of first ligand-receptor pairs would substantially decrease the free energy of formation of later ligand-receptor bonds. Capping phenomena and closely related processes have been observed experimentally,[56,57] and described theoretically.[37] It has been suggested that mutual capping and co-capping always occur during the cell adhesion, unless molecular mobility in the membrane is inhibited.[54] These capping processes may play an important functional role not only in adhesion phenomena, but also can provide a mechanism to allow the stable binding of certain weak ligand-receptor pairs, for example, specific glycosphingolipid-glycosphingolipid interactions or other carbohydrate-carbohydrate recognition and association, that are thereby allowed to participate in intracellular signaling between two contacting cells. In this context it is interesting to note that the role of certain sphingolipid catabolites (sphingosine) as a negative modulator of protein kinase C was proposed.[58,59] A recent study suggests also that sphingosine enhances production of phosphatidic acid, which may trigger intracellular Ca^{2+} mobilization and activate G-protein.[60] The involvement of endogenous ganglioside GM3 in inhibition of epidermal growth factor receptor kinase activity was demonstrated by Weis and Davis.[61] The studies of various glycosphingolipid catabolites show that they may act as an inhibitor and/or activator of protein kinase C and epidermal growth factor receptor kinase.[62,63]

Finally, a large body of evidence has established the role of certain phospholipids, such as diacylglycerol and inositol tri- or diphosphate, as a key component of mitogenic signaling pathways that act by modulating protein kinase C activity and Ca2+ mobilization.[64-66] The observations that some glycosphingolipids and their catabolites affect transmembrane signaling also support the idea that specific glycosphingolipid-glycosphingolipid interactions may occur on the late stages of cellular adhesion to modify intracellular signaling pathway of contacting cells. At least five major plasma membrane glycosphingolipids or phospholipid components, nine identified catabolites or derivatives and a large number of presently unknown compounds may be involved in the control of transmembrane signaling[30] that provides an unique array of a "messenger" molecules for highly specific modulation of intracellular signaling and possibly changes in gene functions following cell-cell type specific adhesion.

The concept of competition between nonspecific repulsion and specific bonding during cell adhesion implies the existence of a repulsive barrier preventing adhesion that must be overcome by bridging molecules, as well as the existence of two phase transitions corresponding, respectively, to the onset of stable cell adhesion and to the onset of maximum cell-cell or cell-substrate contact.[37] Certain configurations of cell-cell contact involving redistribution of receptors into the contact region might be required to achieve a degree of bridging sufficient to overcome the repulsive barrier, and adhesion receptors will become concentrated in regions of cell-cell or cell-substrate contact.

The capping phenomena may also play an important role in the initial stage of cell adhesion as a mechanism which excludes glycocalix components from the site of cell-cell contact. For example, the initial cell-cell contact would be mediated by circulating naturally occurring anticarbohydrate antibodies in order to overcome the size limitation by glycocalyx molecules on the interaction of cell adhesion molecules. (Fig. 1) Then the mutual capping and co-capping processes will initiate the subsequent stages of cellular adhesion with the involvement of selectin-carbohydrate interactions, immunoglobulin-, integrin- and extracellular matrix protein-mediated attachment, carbohydrate- carbohydrate recognition and interactions, gap junction formation and activation of transmembrane and intracellular signaling mechanisms. The experimental evidence clearly shows that BGA-related glycodeterminants may serve as a ligand and/or receptor in cell-cell recognition and adhesion. An important unanswered question is whether or not membrane-associated carbohydrates may function as a fully competent cellular receptor and mediate the internalization and cell uptake of corresponding ligand as well as activate specific intracellular signal transduction pathway?

THE ROLE OF INTEGRINS IN CELL ADHESION

The recent excellent review of Hynes[67] contains an update on this topic, so we will consider here only those aspects of integrin function that are closely related to the aberra-

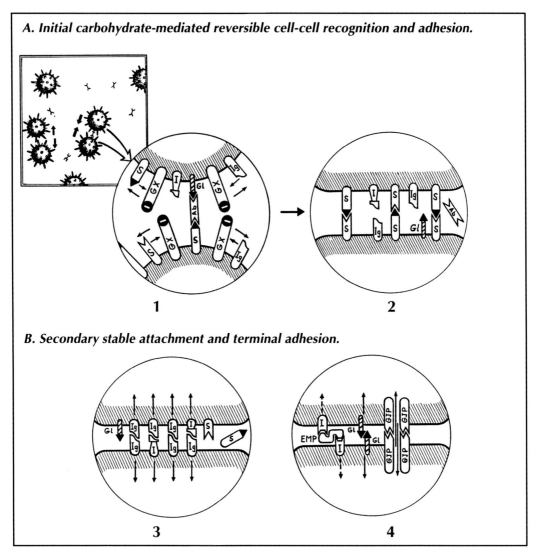

A. Initial carbohydrate-mediated reversible cell-cell recognition and adhesion.

1   **2**

B. Secondary stable attachment and terminal adhesion.

3   **4**

Fig. 1. A model indicating multicomponent and multistep cell-cell type specific recognition and adhesion. The figure summarizes both experimental data and hypotheses discussed in the text. The initial reversible cell-cell recognition (steps A1 & 2) is primarily mediated by specific carbohydrate units (black triangles) expressed on both glycoproteins and glycolipids (G1). This stage involves naturally occurring anticarbohydrate antibodies (Ab), selectins (S), and possibly, glycosphingolipids (G1) and glycoproteins bearing corresponding glycodeterminants. The arrows indicate mutual capping and co-capping processes that presumably may be initiated on reversible stage of cell adhesion. The anticarbohydrate antibody-mediated adhesion is accompanied by subsequent cell membrane component reorganization, activation and translocation of adhesion molecules. This initial event is followed by secondary stable attachment and terminal adhesion which is primarily mediated by protein adhesion systems such as immunoglobulin supergene family of adhesion molecules (Ig), integrins (I), extracellular matrix proteins (EMP), gap junctional proteins (GJP), and probably, by glycosphingolipid-glycosphingolipid (G1) interactions. This stage is accompanied by shedding from the cell surface of primary adhesion molecules such as selectins (S), further clustering, aggregation and cross-linking in the defined zone of morphological cell-cell contact of all essential adhesion receptor-counterreceptor pairs that is required for generation of specific, complete, competent, and adequate intercellular signaling pathways within interacting cells (the arrows on figures B3 & 4). Typically, during the transition period from stage A to stage B of cellular adhesion the specific set of intracellular activating and/or inactivating signals is generated within interacting cells. In some cell type, such as immune cells, specific protein-protein interactions (T cell receptor complex-MHC-peptide complex) are involved on both stage sA and B of cellular adhesion.

tions of cancer cell adhesion. The earlier definition of integrins as fibronectin or vitronecin receptors have proven too restrictive, because there are numerous receptors for each of these glycoproteins and they are not highly specific for individual adhesive ligands.[67] There are now about 20 integrins—the major family of receptor molecules by which cells attach to extracellular matrices and/or to the other cells. Most cells express several integrins and most integrins are expressed on a wide variety of cells. The studies of ligand and adhesive specificity of individual integrins using cell adhesion assay, monoclonal antibodies and affinity chromatography reveal that individual ligands are recognized by more than one integrin and individual integrins can bind to more than one ligand.[67] The molecular transmembrane topography of integrins indicates that the N-terminal domains of α and β subunits combine to form a ligand-binding head of each integrin. Two cytoplasmic domains interact with cytoskeletal proteins and perhaps with other cytoplasmic components. There is some evidence that different integrin subunits may mediate differing cellular responses even to common extracellular adhesion ligands.[68-70] Most integrins interact with the actin-based cytoskeleton, but some of them more likely interacts with intermediate filaments, which are associated with hemidesmosomes.[71-73]

Individual cells can vary their adhesive properties by selective expression of integrins and by modulation of binding properties (affinity and ligand specificity) of integrins. For example, $\alpha_2\beta_1$, integrin on many cells can recognize both collagen and laminin,[74,75] but on platelets it is specific for collagen and not laminin.[76] The specificity and affinity of a given integrin receptor on a given cell are not always constant and numerous examples of modulation of integrin functions include both activation and deactivation of integrins.[67] Often only after cell activation by specific factors the integrins become an effective receptor for corresponding ligands. The effective activation stimuli for β_2 integrins that are expressed on leukocytes vary depending on cell type. Activation can be accomplished by phorbol esters,[77] by adherence to fibronectin[78,79] or by various inflammatory mediators and cytokines

(see below). There is an adhesive cascade leading to attachment of neutrophils and monocytes to endothelial layers. Multiple adhesion and activation steps lead to function of β_2 integrins which provide the necessary adhesive strength, but the specificity comes from the interaction of multiple receptor-counterreceptor pairs and activation steps.[80,81] Strong, antigen-specific adhesion of T lymphocytes to antigen-presenting cells is also mediated by adhesive and activation cascade: weak but specific adhesion via T cell receptor—CD3 complex triggers activation of β_2 integrins leading to strong adhesion.[67,82,83] The activation of β_2 integrins in neutrophils, monocytes and lymphocytes is accomplished by the involvement of protein kinase C in the activation pathways.

The levels of β_1 integins that are widely expressed on lymphocytes and leukocytes[84,85] increase after antigen stimulation. The β_1 integrins mediate attachment of lymphocytes, as well as many other cell types, to extracellular matrix protein and likely play a role in extravasation and migration of activated lymphocytes in tissues during immune responses.[86] Conversion from naive to memory T cells is accompanied by the changes in surface levels of β_1 integrins. The activation mechanisms of β_1 and β_2 integrins in lymphocytes are probably very similar because activation of T cells by antigen or by phorbol esters leads to activation of β_1 integrins and increased adhesion of activtated T cells to collagen, fibronectin and laminin.[87-89]

Phosphorylation of β_1 integrins may cause apparent inactivation of the receptors. During oncogenic transformation by Rous sarcoma virus pp 60 [src] phosphorylates tyrosine residue in the β_1 subunit and this appears to reduce binding of β_1 integrins to both talin and fibronectin.[90] There is additional evidence supporting the correlation of tyrosine phosphorialation of β_1 integrin with inactivation.[91] The specific serine residue in β_1 integrin becomes phosphorylated in mitotic cells and the β_1 integrin from such cells no longer binds to fibronetin, consistent with the rounding and detachment of cells during mitosis.[67] Thus, inactivation of integrins may be regulated by phosphorylation. The constitutively inactive state of the integrins on platelets and leukocytes is crucially important for control of thrombosis and inflammation. Transitorily, during cell migration or mitosis, attached cells need to detach from extracellular matrix or from other cells. In a variety of cell types occupation of integrin receptors by their ligands leads to tyrosine phosphorylation and cytoplasmic alkalinization.[67] It has been suggested the convergence of the signaling pathways triggered by soluble growth factors, extracellular matrix adhesion receptors and tyrosine kinase oncogene products offers a potential explanation for the anchorage dependence of growth of normal cells and its loss in transformed cells.[92] Adhesion of fibroblasts, endothelial cells and lymphocytes to fibronectin causes elevation of cytoplasmic pH, which correlates with the parallel stimulation of spreading and growth.[93-95]

In cases of cytoplasmic alkalinization, as in the case of tyrosine phosphorylation, one can conclude that cell adhesion signals generated by integrins converge and synergize with those triggered by soluble growth factors and oncogenes, because constitutively elevated cytoplasmic pH in transformed cells correlates with anchorage independency of growth.[96] Therefore, in a variety of cell types (platelets, fibroblasts, carcinoma cells, endothelial cells, lymphocytes, monocytes, neutrophils), integrins act as costimulatory signaling receptors and the integrin-mediated signaling pathways converge and synergize with those due to the soluble ligands.[67]

It has been proposed that clustering and cross-linking of integrins brings them together, clustering protein tyrosine kinases in a submembranous patch where they become activated and/or react with their substrates.[67] In an adhesion process integrins provide the strong stable attachment but only after cell activation by other stimuli such as soluble ligands and mediators as well as interactions with extracellular matrix or other cells. The consequences of activation of specific integrins may include a number of different cellular events: a) enhancement of cell adhesion; b) activation of cell proliferation; c) activation of secretion mechanisms; and d) morphological changes, e.g., cytodifferentiation. Particularly interesting is the concept of integrin-mediated anchorage-dependence of growth.[67] Extracellular matrix proteins acting through integrins stimulate tyrosine phosphorylation inside the cells. These signals may converge or synergize with those from growth factor receptors and in transformed cells with oncogene products such as pp 60[src]. The requirement of synergy between soluble growth factors and extracellular matrix adhesion in activation of tyrosine phosphorylation and cell multiplication would give anchorage dependence of growth for normal cells. Subversion or replacement of the need for extracellular matrix generated signals by an oncogene tyrosine kinase activation may lead to anchorage independence of growth for transformed cells. The multiple domains of extracellular matrix molecules cluster different cellular receptors to generate combined intracellular signals analogous to mechanisms generated by clustering receptors in T cells as a result of their interaction with antigen-presenting cells.[67] It has been suggested that the modular structure of extracellular matrix and multiple distinct sites for interaction with cells are designed to cross-link several different surface receptors together in the plane of the membrane, to organize specific cytoskeletal structures and to trigger and/or generate the intracellular signaling.[67] The evidence that fibronectin segments with multiple domains trigger greater tyrosine phosphorylation than do simpler cell-binding domains,[92] and larger fibronection domains promote more cytoskeletal organization[97,98] supports this concept.

Therefore, impairment or subversion of physiological integrin functions in cancer cells may be a key factor contributing significantly to the formation of basic fundamental features of malignant phenotype such as autonomization of cell multiplication, loss of anchorage dependence of growth and aberrant adhesion behavior of cancer cells. (Fig. 2) The crucially important role in physiological function of integrins and other adhesion receptors involve an unique modular structure of extracellular matrix with precise spatial organization of a multiple distinct cell binding sites on extracellular matrix proteins. It seems that this is a requirement for design of

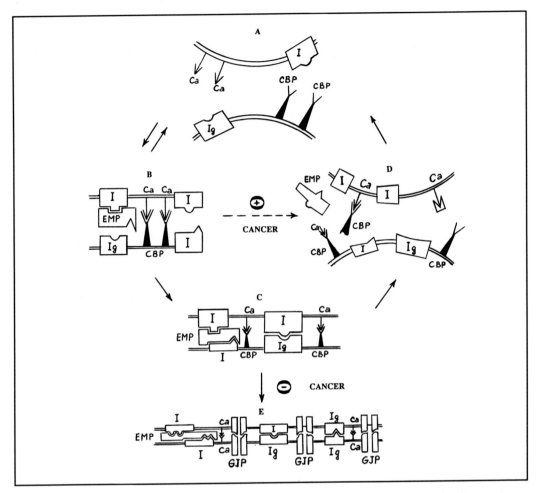

Fig. 2. A model of cell adhesion-cell motility and/or cell multiplication cycle discussed in the text. Stages A,B,C and D comprise reversible cell adhesion/motility/multiplication cycle.

Stages A,B,C, and E comprise irreversible cell adhesion/stable attachment/terminal adhesion cycle.

Stages A and D represent preadhesion or competent state and postadhesion or detachment state of interacting cells, correspondingly. Stages B,C and E indicate the initial reversible adhesion, secondary stable attachment and tissue-specific terminal (irreversible ?) adhesion, correspondingly.

Stage B is primarily mediated by carbohydrate (Ca)- carbohydrate binding protein (CBP) interactions with the involvement of carbohydrate-carbohydrate and integrin (I)- extracellular matrix protein (EMP) interactions in some cellular systems. Stage C is primarily mediated by protein adhesion systems comprising integrin (I), immunoglobulin (Ig), and extracellular matrix protein (EMP) families of adhesion molecules. At stage E the tissue-specific structural proteins become involved and gap junctional intercellular communication channels are opened (GJP-gap junction proteins). Cancer cells lose the ability to display stage E of cell type-specific adhesion..

Stage C, largely dependent upon integrin-mediated functions, gives anchorage dependence for growth of normal cells. Subversion or replacement of the need for extracellular matrix protein generated signals in cancer cells may lead to the anchorage independence of growth and neoplastic cells would pass from stage B to stage D of cell adhesion/motility/multiplication cycle. At stage D (postadhesion or detachment state) multiple antiadhesion mechanisms are displayed: a) dramatic conformational and affinity changes in components of protein adhesion systems (for example, phosphorylation of integrins may cause apparent inactivation of the receptors); b) shedding and/or structural modification of glycodeterminants of cell adhesion as well as shedding of carbohydrate binding proteins, e.g., shedding of selectins occurs at the second stage of leukocyte-endothelial cell adhesion; c) free Ca and CBP would block their cellular counterparts and inhibit the cell-cell or cell-substratum adhesion.

Stage D is followed by cell motility and/or a cell multiplication stage that may restore the adhesion competence of cells (stage A).

the specific cross-link pattern of several different cellular adhesion receptors in the plane of the membrane and subsequent generation of specific intracellular signaling pathway.

The transglutaminase-mediated intra-and intermolecular cross-linking of extracellular matrix proteins may provide and an enzymatic mechanism of supramolecular organization and stabilization of the structure of extracellular matrix. It has been pointed out that primary amine and glucose aldehyde-dependent reactions of covalent modification of primary structure of extracellular matrix proteins may lead to disturbances of the above-named mechanism disorganizing structure and function of extracellular matrix proteins in cancer.[9,99,100] In this context it is interesting to note that the normal morphology of transformed cells can be restored in vitro by addition of fibronectin,[101-103] derived either from normal or malignant cells.[102] On the other hand, associated with neoplastic transformation gross failure in junctional communication would cause uncontrolled cancerous growth.[104]

References

1. Rutishauser, U., Acheson, A., Hall, A.K., Mann, D.M., Sunshine, J. (1989). The neural cell adhesion molecule (NCAM) as a regulator of cell-cell interactions. Science, 240:53-57.

2. Bixby, J.L., Pratt, R.S., Lillien, J., Reichardt, L.F. (1987). Neurite outgrowth on muscle cell surfaces involves extracellular matrix receptors as well as Ca2+ dependent and independent cell adhesion molecules. Proc. Natl. Acad. Sci. USA 84:2555-2559.

3. Lindner, J., Zinser, G., Werz, W., Goridis, C., Bizzini, B., Schachner, M. (1986). Experimental modification of postnatal cerebellar granule cell migration in vitro. Brain Res. 377:298-304

4. Chang, S., Rathjen, F.G., Raper, J.A. (1987). Extension of neurites on axons is impaired by antibodies against specific neural cell surface glycoproteins. J. Cell Biol. 104:355-362.

5. Eggens, I., Fenderson, B., Toyokuni, T., B. Dean, B., Stroud M., Hakomori, S.I. (1989) Specific interaction between Le^x and Le^x determinants: A possible basis for cell recognition in preimplantation embryos and in embryonal carcinoma cells. J. Biol. Chem. 264:9476-9484.

6. Kojima, N., Hakomori., S.I. (1989) Specific interaction between gangliotriaosylceramide (Gg3) and sialosyllactosylceramide (GM3) as a basis for specific cellular recognition between lymphoma and melanoma cells. J. Biol. Chem. 264:20159-20162.

7. Glinsky, G.V. (1990). Immunoselective hypothesis of tumor progression. Role aberrant glycosylation, anti-carbohydrate antibodies, extracellular glyco-macromolecules and glycoamines. J. Tumor Marker Oncology. V.5, N 3, p. 206.

8. Glinsky, G.V. (1990). Glycoamines, aberrant glycosylation and cancer: A new approach to the understanding of molecular mechanism of malignancy. In: Molecular Oncology. Oncodevelopment proteins and clinical applications. XVIIIth meeting of the International Society for Oncodevelopmental Biology and Medicine. Abstract Book, Moscow, USSR, September 23-27, 1990, p.7.

9. Glinsky, G.V. (1992). Glycoamines: Structural-functional characterization of a new class of human tumor markers. In: Serological Cancer Markers. Editor: S. Sell. The Humana Press, Totowa, NJ, Chapter 11, pp. 233-260.

10. Glinsky, G.V., Semyonova-Kobzar, R.A., Berezhnaya, N.M. (1990). Modification of cellular adhesion, metastasizing and immune response by glycoamines: implication in the pathogenetical role and potential therapeutic application in tumoral disease. J. Tumor Marker Oncology., V.5, N 3, p. 231.

11. Glinsky, G.V. (1992). The blood group antigens (BGA)-related glycoepitope. A key structural determinants in immunogenesis and cancer pathogenesis. Critical Reviews in Oncology/Hematology, 12:151-166.

12. Fenderson, B.A., Andrews, P.W., Nudelman, E., Clausen, H., Hakomori, S.I. (1987). Glycolipid core structure switching from globo to lacto-and ganglioseries during retinoic acid-induced differentiation of TERA-2-derived human embryonal carcinoma cells. Dev. Biol. 122, 21-34.

13. Fenderson, B.A., Eddy, E.M., Hakomori, S.I. (1990). Glycoconjugate expression during embryogenesis and its biological significance. BioEssays. 12, 173-79.

14. Lindenberg, S., Sundberg, K., Kimber, S.J., Lundblad, A. (1988). The milk oligosaccharide, lacto-N-fucopentaose l, inhibits attachment of mouse blastocysts on endometrial monolayers. J. Reprod. Fert. 83, 149-158.

15. Springer, G.F. (1984). T and Tn, general carcinoma autoantigens. Science, 224, 1198-1206.

16. Springer, G.F., Cheinsong-Popov, R., Schirrmacher, V., Desoi, P.R., Tegtmeyer, H. (1983). Proposed molecular basis of murine tumor cell-hepatocyte interaction. J.Biol. Chem. 258, 5702-5706.

17. Bird, J.M and S.J. Kimber (1984). Oligosaccharides containing fucose linked α (1-3) and α (1-4) to N-acetylglucosamine cause decompaction of mouse morulae. Dev. Biol., 104:449-460.

18. Clausen, H., Hakomori, S.I. (1989). ABH and related histo-blood group antigens; Immunochemical differences in carrier isotypes and their distribution. Vox Sang. 56, 1-20.

19. Lloyd, K.O. (1988). Blood group antigen expression

in epithelial tumors: influence of secretor status. In: Altered glycosylation in tumor cells. Editors: Ch.L. Reading, S.-I. Hakomori, D.M. Markus, NY. Alan R. Liss, pp. 235-243.

20. Stein, R., Chen, S., Grossman, W., Goldenberg, D.M. (1989). Human lung carcinoma monoclonal antibody specific for the Thomsen-Friedenreich antigen. Cancer Res., 49:32-37.

21. Larsen, E., Palabrica, T., Sajer, S., Gilbert, G.E., Wagner, D.D., Furie, B.C., Furie, B. (1990). PADGEM-dependent adhesion of platelets to monocytes and neutrophils is mediated by a lineage-specific carbohydrate. LNF III (CD15). Cell, 63:467-474.

22. Phillips, M.L., Nudelman, E., Gaeta, F.C.A., Perez, M., Singhal, A.K., Hakomori, S., Paulson, J.C. (1990). ELAM-1 mediates cell adhesion by recognition of a carbohydrate ligand, sialyl-Lex. Science, 250:1130-1132.

23. Polley, M.J., Phillips, M.L., Wagner, E.A., Nudelman, E., Singhal, A.K., Hakomori, S., Paulson, J.C. (1991). CD 62 and endothelial cell-leukocyte adhesion molecule 1 (ELAM-1) recognize the same carbohydrate ligand, sialyl-Lewis X. Proc. Natl. Acad. Sci. USA, 88:6224-6228.

24. Berg, E.L., Robinson, M.K., Mansson, O., Butcher, E.C., Magnani, J.L. (1991). A carbohydrate domain common to both sialyl LeA and SialylLex is recognized by the endothelial cell leukocyte adhesion molecule ELAM-1. J. Biol. Chem., 266:14869-14872.

25. Fukuda, M., Spooncer, E. S., Oates, J.E., Dell, A., Klock, J.C. (1984). Structure of sialylated fucosyl lactosaminoglycan isolated from human granulocytes. J. Biol. Chem, 259:10925-10935

26. Symington, F.W., Hedges, D.L., Hakomori, S.-I. (1985). Glycolipid antigens of human polymorphonuclear neutrophils and the inducible HL-60 myeloid leukemia line. J. Immunol., 134:2498-2506.

27. Picker, L.J., Kishimoto, T.K., Smith, C.W., Warnock, R.A., Butcher, E.C. (1991). ELAM-1 is an adhesion molecule for skin-homing T cells. Nature 349:796-799.

28. Shimizu, Y., Shaw, S., Graber, N., Copal, T.V., Horgan, K.J., VanSeventer, G.A., Newman, W. (1991). Activation-independent binding of human memory T cells to adhesion molecule ELAM-1. Nature 349:799-802.

29. Hakomori, S.-I. (1991). Possible functions of tumor-associated carbohydrate antigens. Current Opinion in Immunology. 3:646-653.

30. Hakomori, S.-I. (1992). Possible new directions in cancer therapy based on aberrant expression of glycosphingolipids: Antiadhesion and ortho-signalling therapy. Cancer Cells (in press).

31. Kojima, N., Hakomori, S. (1992). Synergistic effect of two cell recognition systems: glycosplingolipid-glycosphingolipid interaction and integrin receptor interaction with pericellular matrix protein. Glycobiology (in press).

32. Lawrence, M.B., Springer, T.A. (1991). Leukocytes roll on a selectin at phisiologic flow rates: distinction from and prerequisite for adhesion through integrins. Cell, 65:859-873.

33. Kojima, N., Hakomori, S. (1991). Cell adhesion, spreading and motility of GM3- expressing cells based on glycolipid-glycolipid interaction. J. Biol. Chem 266:17552-17558.

34. Springer, T.A. (1990). Adhesion receptors of the immune system. Nature, 346:425-434.

35. Biberfeld, P., Johansson, A. (1975). Contact areas of cytotoxic lymphocytes and target cells. Exp. Cell Res. 94:79-87.

36. Verschueren, H. (1985) Interference reflection microscopy in cell biology: methodology and applications. J. Cell Sci. 75:279-301.

37. Bell, G.I., Dembo, M., Bongrand, P. (1984). Cell adhesion. Competition between nonspecific repulsion and specific bonding. Biophys. J. 45:1051-1064.

38. Williams, A.F., Barclay, A. (1986). In Immunochemistry (Eds. Weir, D.M., Herzenberg, L.A., Blackwell, C., Herzenberg, L.A.) Ch. 22, Blackwell, Oxford.

39. Kaizu, T., Levery, S.B., Nudelman, E., Stenkamp, R.E., Hakomori, S. (1986). Novel fucolipids of human adenocarcinoma: monoclonal anitbody specific for trifucosyl LeY (111^3 FucV^3 FucVI^2 FucnLc$_6$) and a possible three-dimensional epitope structure. J. Biol. Chem. 261:11254-11258.

40. Hakomori, S. (1986). Glycosphingolipids. Sci. Am. 254:44-53.

41. Hakomori, S. (1989). Aberrant glycosylation in tumors and tumor-associated carbohydrate antigens. Adv. Cancer Res. 52:257-331.

42. Picker, L.J., Warnock, R.A., Burns, A., Doerschuk, C.M., Berg, E.L., Butcher, E.C. (1991). The neutrophil selectin LECAM-1 presents carbohydrate ligands to the vascular selectins ELAM-1 and GMP-140. Cell, 66:921-933.

43. VonAndrian, U.H., Chambers, J.D., McEvoy, L.M., Bargatze, R.F., Arfors, K.-E., Butcher, E.C. (1991). Two-step model of leukocyte-endothelial cell interaction in inflammation: distinct roles for LECAM-1 and the leukocyte b2 integrins in vivo. Proc. Natl. Acad. Sci. USA 88:7538-7542.

44. Smith, C.W., Kishimoto, T.K., Abbass, O., Hughes, B., Rothlein, R., McIntire, L.V., Butcher, E., Anderson, D.C. (1991). Chemotactic factors regulate lectin adhesion molecule 1 (LECAM-1)-dependent neutrophil adhesion to cytokine-stimulated endothelial cells in vitro. J. Clin. Invest. 87:609-618.

45. Kishimoto, T.K., Warnock, R.A., Jutila, M.A., Butcher, E.C., Lane, C., Anderson, D.C., Smith, C.W. (1991). Antibodies against human neutrophil LECAM-1 (LAM-1/Leu-8/DREG-56 antigen) and endothelial cell ELAM-1 inhibit a common CD-18-independent adhesion pathway in vitro. Blood 78:805-811.

46. Arfors, K.E., Lundberg, C., Lindbom, L., Lundberg, K., Beatty, P.G., Harlan, J.M. (1987). A monoclonal antibody to the membrane glycoprotein complex

CD18 inhibits polymorphonuclear leukocyte accumulation and plasma leakage in vivo. Blood 69:338-340.

47. Kishimoto, T.K., Larson,R.S., Corbi, A.L., Dustin, M.L., Staunton, D.E., Springer, T.A. (1989a). The leukocyte integrins. Adv. Immunol. 46:149-182.

48. Carlos, T.M., Harlan, J.M. (1990). Membrane proteins involved in phagocyte adherence to endothelium. Immunol. Rev. 114:5-28.

49. Lawrence,M.B.,Smith,C.W.,Eskin,S.G.,McIntire, L.V. (1990). Effect of venous shear stress on CD18-mediated neutrophil adhesion to cultured endothelium. Blood 75:227-237.

50. Kishimoto, T.K, Jutila, M.A., Berg, E.L., Butcher, E.C. (1989b). Neutrophil Mac-1 and MEL-14 adhesion proteins inversely regulated by chemotactic factors. Science 245:1238-1241.

51. Berg, M., James, S.P. (1990). Human neutrophil release the Leu-8 lymph node homing receptor during cell activation. Blood 76:2381-2388.

52. Griffin, J.D., Spertini, O., Ernst, T.J., Belvin, M.P., Levine, H.B., Kanakura, Y., Tedder,T.F. (1990). Granulocyte-macrophage colony-stimulating factor and other cytokines regulate surface expression of the leukocyte adhesion molecule-1 on human neutrophils, monocytes and their precursors. J. Immunol. 145:576-584.

53. Singer, S.J., Kupfer, A. (1988). In: The T-cell receptor. M.M. Davis and J. Kappler, Eds. Alan R. Liss, New York, pp. 361-376.

54. Singer, S.J. (1992). Intercellular communication and cell-cell adhesion. Science, 255:1671-1677.

55. Singer, S.J. (1976). In: Surface membrane receptors, interface between cells and their environment. R.A. Bradshaw, W.A. Frazier, R.C. Merrell, D.I. Gottlieb, R.A. Hogue - Angeletti, Eds. Plenum Press, New York, pp.1-24.

56. Weis, R.M., Balakrishnan, K., Smith, B.A., McConnell, H.M. (1982). Stimulation of fluorescence in a small contact region between rat basophil leukemia cells and a planar lipid membrane targets by coherent evanescent radiation. J. Biol. Chem. 257:6440-6445.

57. McCloskey, M.A., Poo, M.-M. (1986). Contact-induced redistribution of specific membrane components: local accumulation and development of adhesion. J. Cell Biol. 102:2185-2196.

58. Hannun,Y.A.,Bell,R.M.(1987).Lysosphingolipids inhibit protein kinase C: Implication for the sphingolipidoses. Science, 235:670-674.

59. Hannun, Y.A., Bell, R.M. (1989). Functions of sphingolipids and sphingolipid breakdown products in cellular regulation. Science, 243:500-507.

60. Zhang, H., Buckely, N.E., Gibson, K., Spiegel, S. (1990). Sphingosine stimulates cellular proliferation via protein kinase C-independent pathway. J. Biol. Chem. 265:76-81.

61. Weis, F.M.B., Davis, R.J. (1990). Regulation of epidermal growth factor receptor signal transduction: Role of gangliosides. J. Biol. Chem. 265:12059-12066.

62. Hanai, N., Dohi, T., Nores, G.A., Hakomori, S. (1988). A novel ganglioside, de-N-acetyl-GM3 (II^3NeuNH_2Lac Cer), acting as a strong promoter for epidermal growth factor receptor kinase and as a stimulator for cell growth. J. Biol. Chem. 263:6296-6301.

63. Igarashi, Y., Nojiri, H. Hanai, N., Hakomori, S. (1989). Gangliosides that modulate membrane protein function. Meth. Enzymol., 179:521-541.

64. Nishizuka, Y. (1984). The role of protein kinase C in cell surface signal transduction and tumor promotion. Nature, 308:693-698.

65. Berridge, M.J. (1983). Rapid accumulation of inositol triphosphate reveals that agonists hydrolyse polyphosphoinositides instead of phosphatidyl-inositol. Biochem. J. 212:849-858.

66. Exton, J.H. (1990). Singaling through phosphatidylcholine breakdown. J. Biol. Chem. 265:1-4.

67. Hynes, R.O. (1992). Integrins: versatility, modulation and signaling in cell adhesion. Cell, 69:11-25.

68. Elices, M.J., Urry, L.A., Hemler, M.E. (1991). Receptor functions for the integrin VLA-3: fibronectin, collagen and laminin binding are differentially influenced by Arg-Gly-Asp peptide and by divalent cations. J. Cell Biol. 112:169-181.

69. Wayner, E.A., Orlando, R.A., Cheresh, D.A. (1991). Integrin a_vb_3 and a_vb_5 contribute to cell attachment to vitronectin but differentially distribute on the cell surface. J. Cell Biol., 113:919-929.

70. Chan, B.M.C., Kassner, P.D., Schiro, J.A., Byers, R., Kupper, T.S., Hemler, M.E. (1992b). Distinct cellular functions mediated by different VLA integrin a subunit cytoplasmic domains. Cell, 68:1051-1060.

71. Stepp, M.A., Spurr-Michaud, S., Tisdale, A., Elwell, J., Gipson, I.K. (1990). α_6/ β_4 integrin heterodimer is a component of hemidesmosomes. Proc. Natl. Acad. Sci. USA 87:8970-8974.

72. Sonnenberg, A., Calafat, J., Janssen, H., Daams, H., van der Raaij-Helmer, L.M., Falcioni, R., Kennel, S.J., Aplin, J.D., Baker, J., Loizidou, M., Garrod, D. (1991). Integrin α_6/ β_4 complex is located in hemidesmosomes, suggesting a major role in epidermal cell-basement membrane adhesion. J. Cell Biol. 113:907-917.

73. Kurpakus, M.A., Quaranta, V., Jones,J.C.R.(1991). Surface relocation of $alpha_6$/$beta_4$ integrins and assembly of hemidesmosomes in an in vitro model of wound healing. J. Cell Biol. 115:1737-1750.

74. Elices, M.J., Hemler, M.E. (1989). The human integrin VLA-2 is a collagen receptor on some cells and a collagen/laminin receptor on others. Proc. Natl. Acad. Sci. USA 86:9906-9910.

75. Kirchhofer, D., Languino, L.R., Ruoslahti, E., Pierschbacher, M.D. (1990b). $\alpha_2\beta_1$ integrins from different cell types show different binding specificities, J. Biol. Chem. 265:615-618.

76. Staatz, W.D., Rajpara, S.M., Wayner, E.A., Carter, W.G., Santoro, S.A. (1989). The membrane glycoprotein la-lla (VLA-2) complex mediates the Mg^{++}-dependent adhesion of platelets to collagen. J. Cell Biol. 108: 1917-1924.

77. Wright, S.D., Silverstein, S.C. (1982). Tumor-promoting phorbol esters stimulate C3b and C3bi receptor-mediated phagocytosis in cultured human monocytes. J. Exp. Med. 156:1149-1164.

78. Wright, S.D., Rao, P.E., Van Voorhis, W.C., Craigmyl, L.S., Jida, K., Talle, M.A., Westberg, E.F., Goldstein, G., Silverstein, S.C. (1983). Identification of the C3bi receptor on human monocytes and macrophages by using monoclonal antibodies. Proc. Natl. Acad. Sci. USA 80:5699-5703.

79. Wright, S.D., Licht, M.R., Craigmyl, L.S., Silverstein, S.C. (1984). Communication between receptors for different ligands on a single cell: Ligation of fibronectin receptors induces a reversible alteration in the function of complement receptors on cultured human monocytes. J. Cell Biol. 99:336-339.

80. Butcher, E.C. (1991). Leukocyte-endothelial cell recognition: three (or more) steps to specificity and diversity. Cell 67: 1033-1036.

81. Hynes, R.O., Lander, A.D. (1992). Contact and adhesive specificities in the association, migrations and targeting of cells and axons. Cell 68:303-322.

82. Dustin, M.L., Springer, T.A. (1991). Role of lymphocyte adhesion receptors in transient interactions and cell locomotion. Annu. Rev. Immunol. 9:2-66.

83. van Kooyk, Y., van de Wiel-van Kamenade, P., Weder, P., Kuijpers, T.W., Figdor, C.G. (1989). Enhancement of LFA-1-mediated cell adhesion by triggering through CD2 or CD3 on T lymphocytes. Nature 342:811-813.

84. Hemler, M.E. (1990). VLA proteins in the integrin family: structures, functions and their role on leukocytes. Annu. Rev. Immunol. 8:365-400.

85. Shimizu, Y., Shaw, S. (1991). Lymphocyte interactions with extracellular matrix. FASEB J. 5:2292-2299.

86. Ferguson, T.A., Mizutani, H., Kupper, T.S. (1991). Two integrin-binding peptides abrogate T cell-mediated immune responses in vivo. Proc. Natl. Acad. Sci. USA 88:8072-8076.

87. Shimizu, Y., van Seventer, G., Horgan, K.J., Shaw, S. (1990b). Regulated expression and binding of three VLA (β 1) integrin receptors on T cells. Nature 345:250-253.

88. Chan, B.M., Wong, J.G., Rao, A., Hemler, M.E. (1991). T cell receptor-dependent, antigen-specific stimulation of a murine T cell clone induces a transient, VLA protein-mediated binding to extracellular matrix. J. Immunol. 147:398-404.

89. Wilkins, J.A., Stupack, D., Stewart, S., Caixia, S. (1991). β₁ integrin-mediated lymphocyte adherence to extracellular matrix is enhanced by phorbol ester treatment. Eur. J. Immunol. 21:517-522.

90. Tapley, P., Horwitz, A.F., Buck, C.A., Burridge, K., Duggan, K., Hirst, R., Rohrschneider, L. (1989). Analysis of the avain fibronectin receptor (integrin) as direct substrate for pp60[v-src]. Oncogene 4:325-333.

91. Horvath, A.R., Elmore, M.A., Kellie, S. (1990). Differential tyrosine-specific phosphorylation in integrin in Rous sarcoma virus transformed cells with differing transformed phenotypes. Oncogene 5:1349-1357.

92. Guan, J.L., Trevithick, J.E., Hynes, R.O. (1991). Fibronectin/integrin interaction induces tyrosine phosphorylation of a 120kDa protein. Cell Regul. 2:951-964.

93. Schwartz, M.A., Both, G., Lechene, C. (1989). Effect of cell spreading on cytoplasmic pH in normal and transformed fibroblasts. Proc. Natl. Acad. Sci. USA 86:4525-4529.

94. Schwartz, M.A., Ingber, D.E., Lawrence, M., Springer, T.A., Lechene, C. (1991). Multiple integrins share the ability to induce elevation of intracellular pH. Exp. Cell Res. 195:533-535.

95. Ingber, D.E., Prusty, D., Frangioni, J.V., Cragoe, E.J., Jr., Lechene, C., Schwartz, M.A. (1990). Control of intracellular pH and growth by fibronectin in capillary endothelia cells. J. Cell Biol. 110:1803-1811.

96. Schwartz, M.A., Rupp, E.E., Frangioni, J.V., Lechene, C.P. (1990). Cytoplasmic pH and anchorage-independent growth induced by v-Ki-ras, v-src or polyoma middle T. Oncogene 5:55-58.

97. Obara, M., Kang, M.S., Yamada, K.M. (1988). Site-directed mutagenesis of the cell-binding domain of human fibronectin: separable, synergistic sites mediate adhesive function. Cell 53:649-657.

98. Woods, A., Johansson, S., Hook, M. (1988). Fibronectin fibril formation involves cell interactions with two fibronectin domains. Exp. Cell Res. 177:272-283.

99. Glinsky, G.V. (1989). Glycoamines: Biochemistry of a new class of humoral tumor markers. J. Tumor Marker Oncology, 4:193-221.

100. Glinsky, G.V., Surgova, T.M., Sidorenko, M.V., Vinnitsky, V.B., Kolesnik, L.A., Shvachko, L.A. (1990). Interaction of glycoamines and polyamines within the system of α₂ macroglobulin-dependent regulation of plasma transglutaminase. A possible molecular mechanism of invasion and implantation in embryogenesis and cancer. J. Tumor Marker Oncology, 5:161-166.

101. Yamada, K.M., Yamada, S.S., Pastan, I. (1976). Cell surface protein partially restores morphology, adhesiveness, and contact inhibition of movement to transformed fibroblasts. Proc. Natl. Acad. Sci. USA, 73:1217-1221.

102. Hayman, E.G., Engwall, E., Ruoslahti, E. (1981). Concomitant loss of cell surface fibronectin and laminin from transformed rat kidney cells. J. Cell Biol., 88:352-357.

103. Ali, I.U., Mautner, V.M., Lanza, R.P., Hynes, R.O. (1977). Restoration of normal morphology, adhesion and cytoskeleton in transformed cells by addition of a transformation-sensitive surface protein. Cell, 11:115-126.

104. Loewenstein, W.R. (1979). Junctional intercellular communication and the control of growth. Biochim. Biophys. Acta, 560:1-65.

CHAPTER 3

Cancer Cell Adhesion:

Preservation of Recognition Function and Impairment of Stable Attachment, Tissue-Specific Terminal Adhesion, Intracellular Signaling and Adequate Cellular Response

The extracellular matrix regulates cell adhesion, cell migration, morphogenesis, and participates in control of cell proliferation and differentiation. It is composed of several glycomacromolecules such as collagens, fibronectin, laminin, and various proteoglycan and glycoproteins.[1-4] The composition of the extracellular matrix of each cell type varies, and the matrix of cancer cells is more sparse and disorganized than matrix of their normal counterparts because some of the transformed cells produce less extracellular matrix component, or the enhanced proteolytic activity of malignant cells causes the impairment in extracellular matrix formation.[5-8] Cell interaction with fibronectin-containing matrices is critical for embryonic development, cytodifferentiation, wound healing and cancer metastasis.[9-16] Fibronectin and fibronectin receptors are co-distributed in a fibrillar array on cultured fibroblasts.[17,18] Furthermore, the cell surface distribution of fibronectin receptors is related to cell behavior. In stationary fibroblasts, the fibronectin receptor is highly organized and distributed identically to extracellular matrix fibronecin fibrils. In highly migratory neural crest cells and embryonic somatic fibroblasts this organization is lost and the fibronectin receptor appears diffuse.[12] Similarly, the oncogenic transformation typically leads to disorganization of the fibronectin receptors and loss of fibronectin matrix.[18,19] The diminished fibronectin matrix in most transformed cells may be because of deficient fibronectin synthesis, cell surface binding, assembly, or fibronectin proteolysis.[20-23]

It has been suggested that phosphorylation of the fibronectin receptor β1 subunit in cells transformed by viruses encoding tyrosine kinases is responsible for the impaired fibronectin receptor function and abnormal distribution of the fibronectin receptors.[24] The evidence has been presented that the fibronectin receptor is organized in fibrils on the cell surface in response to extracellular fibronectin,[18] and the fibronectin receptor is required for optimal fibronectin matrix assembly.[18,25] The diffuse fibronectin receptor phenotype of the SV-40-transformed cells appears to be related to loss of the fibronectin matrix rather than to impaired fibronectin receptor function, because exogenous fibronectin restores fibronectin matrix and receptor organization of transformed cells.[18] Similarly, other cells lacking a fibronectin-containing pericellular matrix also lack organized fibronectin receptors that emphasize the existence of related systems controlling fibronectin deposition and recognition by receptor-bearing cells.[18] The profound change in fibronectin receptor organization associated with fibronectin matrix recognition was early evidence suggesting a major role of the fibronectin receptor in transmission of information about the state of the extracellular matrix to the cells. The alterations in fibronectin receptor function may dictate cell behavior, e.g., by phosphorylation of the fibronectin receptor inducing decreased fibronectin binding and hence increased cell motility.[12] The fibronectin receptor system may serve as a signal transducer between the external fibronectin matrix and the cell, regulating adhesive interaction and motility,[11] cytodifferentiation,[26,27] growth promotion[28] and the regulation of gene expression.[29]

Two major isotypes of fibronectin have been known: One is found in plasma and the other is present in extracellular matrix or is secreted in the cultured medium of fibroblasts.[30,31] Fibronectin extracted from normal adult tissues has similar properties as plasma fibronectin. In contrast, fibronectin isolated from various fetal connective tissues, amniotic fluid, placenta as well as from various types of human carcinoma and sarcoma and their cell lines were found to have the immunochemical similarity to pericellular matrix fibronectin.[32] It has been found that co-presence of one amino acid residue, tyrosine, and the O-glycosylation at threonine (NeuAcα 2-3 Gal β1-3 GalNac; Gal β1-3 GalNac or GalNac) is essential to form a new specific conformational epitope common to the pericellular matrix fibronectin, embryonal and cancer type fibronectins.[33] Since this epitope contains T-and Tn-antigens related sequences it may be involved in cell adhesion process. Both plasma and extracellular matrix human and hamster fibronectin has been shown to contain blood group T-antigen.[34] The O-glycosylation plays an important role in creation of specific conformational epitopes with unique immunochemical specificity. Blood group M and N specificity was found to be determined by the terminal pentapeptide sequences (Ser-Ser-Thr-Thr-Gly and Leu-Ser-Thr-Thr-Glu, correspondingly) of glycophorin with glycosylation at the serine and subsequent two threonine residues. The glycosylation in these cases is essential to maintain the active epitope conformation for M and N blood group antigen specificity.[35,36] In all these cases the biospecifity is maintained by combination of specific animo acid sequence and a defined glycosylation, and neither the carbohydrate structure nor the amino acid sequence per se may be unique.[33]

Temporally and spatially directed specific cell migrations are very important during development and metastasis.[37,38] In the embryo cells migrate in association with both extracellular matrix and the basal lamina secreted by epithelia. Fibronectin is a principle component of the extracellular matrix and has been shown to support cell migration in vitro and in vivo.[9,39,40] It has been observed by the use of blocking antibodies and competitive peptides that inhibit cell migration when injected into the embryo. The presence of a high concentration of G_{D2} and G_{D3} gangliosides in the melanoma cell matrix, which is the main site of cell adhesion, infiltration, and invasion, indicate the possible role of specific carbohydrate determinants in those processes.[41] Furthermore, attachment of melanoma cells to the extracellular matrix proteins (fibronectin) was described to be mediated by G_{D3} ganglioside.[42] Laminin is the predominant, noncollagenous component of basal lamina that mediates cell migration on the basal lamina.[43] Laminin is present in areas of cell migration in vivo,[40] and purified laminin supports cell migration in vitro.[43] A cell surface laminin receptor of 68 KD is involved in cell attachment to laminin,[44] and antireceptor antibody inhibits carcinoma cell binding to laminin.[45] The cell surface enzyme $\beta_{1,4}$ galactosyltransferase is one laminin receptor which participates during cell migration on basal lamina matrices by binding to specific N-linked oligosaccharide substrates within laminin molecules.[43,46] Galactosyltransferase functions as a receptor during cell migration on basal lamina by interacting with specific N-linked oligosaccharide residues in laminin, but it does not affect initial cell attachment to laminin.[46] Reagents that inhibit galactosyltransferase binding to its laminin substrate also inhibit cell spreading and migration on laminin, but do not affect initial cell adhesion to laminin. Consuming potential galactosyltransferase binding sites on laminin by prior glycosylation does not inhibit cell attachment to laminin, but totally inhibits cell spreading and migration.[46] Thus, initial adhesion to laminin may be mediated by the binding of the 68KD and integrin receptors to their peptide ligands,[47] while cell spreading and subsequent migration requires surface galactosylransferase binding to its appropriate N-linked oligosaccharide substrate.[46] Mutual capping and co-capping processes may be responsible for laminin-induced vectoral insertion of galactosyltransferase into growing lamellipodia. It has been shown that laminin-containing matrix (but not fibronectin-containing one) induces the stable expression of galactosyltransferase onto cell lamellipodia and filopodia where it mediates subsequent cell spreading and migration.[48] Cell migration requires the dissociation of cells from the matrix. The galactosyltransferase offers a potential catalytic mechanism whereby cell adhesion to the substrate could be released if a suitable galactose donor were available to the surface enzyme.[48,49] The attached GlcNac galactose would then enable the enzyme (and the cell) to dissociate from its substrate. Combined action of galactosidase and galactosyltransferase would provide a specific regulatory mechanism controlling carbohydrate-mediated cell attachment to and dissociation from extracellular matrix during cell spreading and migration. Obviously, the proteolytic release and/or shedding of adhesive membrane components are alternative possibilities. (Fig. 1)

Therefore, two major extracellular adhesion proteins (fibronectin and laminin) both contain biologically active peptide cell adhesion epitopes and specific carbohydrate structures mediating cellular recognition and adhesion. The difference in strength of cell adhesion, mediated by peptide and carbohydrate cell adhesion epitopes, and concomitant

conformational and adhesion affinity changes, that follow carbohydrate- or protein- dependent cell contact, may be the important factors generating locomotive forces during cell migration. (Fig. 1)

The key features of cancer cell adhesion are: a) preservation of cell recognition function and the initial reversible steps of cell-cell or cell-substrate adhesion; and b) impairment of the ability to display secondary strong adhesion and terminal tissue specific cell-cell and cell-substrate contacts. It has been suggested that associated with neoplastic transformation gross failure in junctional communication would cause uncontrolled cancerous growth.[50] In the chapter describing the integrin role in cell adhesion it has been pointed out that extracellular matrix protein-generated cross-linking of several different adhesion receptors in the plane of the membrane is critically important to organize specific cytoskeletal structures and to trigger and/or generate intracellular signaling. It seems that abrogation of clusterization and cross-linking of adhesion receptors in the membrane of cancer cells may play an important role in dysfunction of cellular adhesion in transformed cells. It is well known that intracellular transglutaminase participates in cross-linking of cellular receptors and cytoskeletal component and that malignant transformation is accompanied by dramatic loss of activity of intracellular transglutaminase in cancer cells. Therefore, one of the main factors that causes the impairment of strong cellular adhesion, adequate intracellular signaling and cellular response in cancer may be the loss of intracellular transglutaminase activity and subsequent abolition of cell adhesion receptor cross-linking function in malignant cells.

The extracellular (plasma) transglutaminase may also play the critical role. One of the key enzymes involved in the formation of extracellular matrix and in the control of adhesive properties of cells is the plasma transglutaminase (factor XIII a). The zymogen form of plasma transglutaminase (factor XIII) undergoes thrombin-induced proteolytic activation before it may catalyze γ-glutamyl-Σ-lysine inter- and intramolecular cross-linking.[51,52] Transglutaminase also catalyzes the

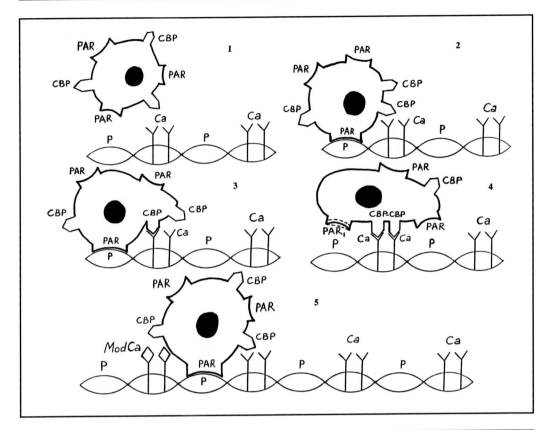

Fig. 1. A model of cell motility based on sequential action of peptide and carbohydrate cell adhesion epitopes (see text).

EMP-   extracellular matrix protien(s), (mainly fibronectin).
BLP-   basal lamina protein(s),(mainly laminin and collagen).
P-   peptide cell adhesion determinant(s).
Ca-   carbohydrate cell adhesion determinant(s).
PAR-   peptide adhesion receptor(s), (integrins, immunoglobulis, etc.).
PAR.-   nonactive peptide adhesion receptor(s).
CBP-   carbohydrate binding protein(s), (cell surface lectins, glycosyltransferases, etc.).
ModCa-   glycodeterminant(s) of cell adhesion modified by glycosyltransferases (may serve as adhesion ligand for other type of cells or for the same cells after action of glycosidases).
CBP-CBP-   cluster of carbohydrate-binding proteins.
1,2,3,4,5- sequential stages of cell motility based on carbohydrate-mediated haptotaxis.

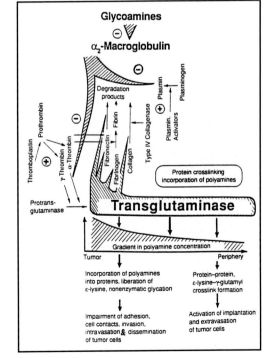

Fig. 2. "Cooperative" mechanism of the antiadhesive action of glycoamines and polyamines.[62]

competitive reaction of the incorporation of primary amines (particularly, the diamine putrescine) into proteins and thereby inhibits the formation of γ-glutamyl-Σ-lysine inter- and intramolecular cross-links.[53,54] The substrates for plasma transglutaminase are fibrin, cell adhesion proteins (fibronectin, collagen)[55-59] and protease inhibitors (α2 macroglobulin).[60] Plasma transglutaminase catalyzes the cross-linking of fibronectin and collagen, and fibronectin and fibrin, contributing in this way to the formation of extracellular matrix.[55-59] It has been suggested[61-63] that in cancer, under the condition of hyperpolyaminenia and hyperglycoaminemia, the restraint of the inhibitory action of α2 macroglobulin brings about the thrombin-mediated activation of plasma transglutaminase and increases the activity of the cascade of proteolytic enzymes. (Fig. 2) In the zone of increased concentration of polyamines the activated transglutaminase incorporates polyamines into proteins thereby preventing γ-glutamyl-Σ-lysine protein-protein cross-linking with the liberation of lysine Σ-amino groups in proteins for ubiquitination (intracellular proteins) and/or nonenzymatic glycation. The attachment of ubiquitin to the proteins activates their proteolysis, and nonenzymatic glycation also alters the sensitivity of proteins to the proteolysis.[64,65] It has been reported that nonenzymatic glycation decreases the affinity of the interaction between fibronectin and collagen.[66] Thus, the concentration gradient of the primary amines between proliferating and nonproliferating tissues and primary amine-dependent restraint of the activity of protease inhibitors may trigger the biochemical mechanism of invasion and implantation.[61-63]

Malignant Behavior of Cancer Cells: Biological Similarities to the Embryonal and Specialized Normal Cells

Individual cells, cell groups and entire tissues move in the embryo at particular sites, and at precisely defined stages of development, traverse well-defined pathways and localize finally at particular sites of the body in specific association with other tissues often at sites quite distant from their original loca-

tions.[67-69] Tissue and cell movement ceases after establishment of final tissue and/or organ organization, except in some pathological and pathophysiological conditions such as inflammation, wound repair, regeneration, immunogenesis and cancer. It has been shown[70,71] that tissue invasion and metastatic dissemination occur during embryonic development as elements of normal morphogenesis. In this context the processes are particularly important that stabilize and preserve the organization of tissues and organs following the events of cell movement during morphogenesis.[38,70,71] The initiation of cell migration involves a reduction of mutual cellular adhesion and a loss of junctional contacts with adjacent cells.[72-74] During cell migration the temporary adhesive interaction of cells with the substratum is necessary to generate locomotive traction. In culture, cells tend to migrate preferentially from sites to which they adhere relatively poorly to sites of stonger adhesion.[75-78]

The haptotaxis is the guidance of cell migration over planar surfaces by spatially distributed differences in the adhesive character of the substratum, with cells migrating toward regions of higher adhesiveness.[75] It has been suggested that haptotactic behavior on a morphological level is a consequence of competition between pseudopods: presumably those pseudopods contacting a more adherent area of the substratum will be less readily distracted than those pseudopods adherent at the same time to less adherent sites.[79] Cell adhesion glycoepitope-mediated haptotactic behavior may be the major factor determining locomotive activity of cancer cells and directing their migratory pathway.

Adhesive interaction of cell with cell and cell with extracellular matrix is essential for the mechanical contiguity of solid tissues and plays an important role in the organization of coherent tissues.[38] It has been proposed that the tissue affinities displayed in different experimental systems in vitro play an important role in regulating the patterning of tissues in vivo.[80-82] The differential adhesion hypothesis represents the analysis of this cellular behavior and generalized the concept that cell migration is regulated by haptotactic interactions of cell with cell and cell with substratum.[83-88] The carbohydrate mediated, particularly BGA-related glycoepitope-dependent, differential homotypic and heterotypic adhesion of cancer cell may be the major factor contributing to the metastatic dissemination of malignant cells. It has been suggested that differential adhesion of transformed cells may play an important role in cancer invasion and metastatic dissemination.[89]

In embryogenesis populations of migrating cells show vectorial dispersal from sites of origin to sites in the periphery. The phenomenon of contact inhibition of cell motility has been suggested as an explanation for directional migration.[90-92] Contact inhibition of cell motility has been defined as the cessation of continued cell locomotion in a direction which has resulted in a collision with a second cell.[90,92] One manifestation of contact inhibition of cell locomotion is contact paralysis, a cessation of pseudopidial protrusion at sites of contact with other cells.[91,93-95] Acording to this concept, inability of cancer cells to convert initial labile carbohydrate-mediated cell adhesion into secondary stable attachment causes the impairment of contact inhibition.

The basal lamina is the specialized element of extracellular matrix that separates the basal surface of epithelia from the connective tissue and interstitial matrix below.[96] The mechanical rupture and/or enzymatic digestion are presumably the mechanisms that permit penetration of the basal lamina by embryonic cells (an invasion). Corresponding proteolytic enzymatic activities have been demonstrated for capillary endothelium[97] and the lymphocyte precursor cells that invade the lymphopoietic organs.[98] Similarly, invasive tumor cell secrete a variety of degradative enzymes.[99-101]

The major components of the basement membrane are common to all basement membrane studied: collagen IV, laminin and heparan sulfate proteoglycan.[102] Tumor cells need to attach to basement membrane in order to degrade this structure.[103,104] It has been shown that tumor cells bind to collagen IV and laminin[105] and binding initiates subsequent events thereby inducing the production of a cascade of degradative enzymes.[100,106]

Binding of tumor cells to laminin induces the production of collagenase IV by the cell.[107] Thus, invasion of cancer cells through basement membrane begins with tumor cell attachment to its surface thereby activating the extracellular enzymathic cascade (plasminogen activator-plasminogen-plasmin-procollagenase IV-collagenase IV) required for the production of collagenase IV and the degradation of basement membrane collagen.[108-110] The close proximity of the cell surface to the basement membrane may facilitate the degradative process by localizing the enzymes and excluding large molecular weight protease inhibitors such as $\alpha 2$ macroglobulin. Receptors on the tumor cell surface have been suggested to target the proteases to the invasive front and specific protease amino acid sequences have been proposed to bind these receptors.[111] Several proteases, like plasmin, collagenases, elastase and cathepsins, participate in the degradation of the extracellular matrix.[7,112] The involvement of catalytically active plasminogen activator in in vitro and in vivo invasion has been indicated.[113,114]

Additionally, tumor cells secrete specific proteases that are able to degrade some of the matrix component[23] such as fibronectin. Furthermore, transfection experiments have indicated a direct role for certain proteases in the invasion of tumor cells.[115] Since high proteolytic activity is usually associated with malignant or proliferating cells, which frequently secrete growth-promoting polypeptides, a possible explanation is that cancer cells produce autocrine growth factors which regulate their proteolytic activity.[106] There are a number of indications that in fact several growth factors have an ability to modulate the expression and activity of proteases and their inhibitors. For instance transforming growth factor α and epidermal growth factor enhance the expression of plasminogen activators[116,117] and of transin[118] in different types of cultured cells. Fibroblast growth factor and transforming growth factor β both induce the expression and the production of both plasminogen activator and endothelial type (type 1) plasminogen activator inhibitor.[106] Thus, autocrine secretion of polypeptide growth factors by cancer cells is evidently the mechanism that adds to the invasive properties of cancer cells through autoactivation of proteolytic enzymes and disturbances of their inhibitors.[106]

As we already have mentioned laminin appears to mediate the interactions of many cells with basement membranes by promoting their attachment, proliferation and differentiation.[119] It has been suggested that the high affinity of metastatic cells to laminin favors their binding to basement membranes and that the interaction between laminin and laminin receptors triggers the production of the basement membrane degradative enzymes.[107] One of the major cell attachment sites in laminin has been identified as a pentapeptide YIGSR.[47] Coated onto plastic petri dishes, synthetic YIGSR containing peptides support the attachment of MCF-7 breast carcinoma cells, CHO cells, PAM212-epidermal cells, B16 melanoma cells and HT1080 fibrosarcoma cells.[47] If laminin is used as a substrate, the YIGSR peptide inhibits cell attachment to it. Several lines of evidence suggest that attachment to laminin is critical step in metastasis.[109] Highly metastatic cells have an increased number of laminin receptors and show greater laminin binding than their benign counterparts.[120,121] Tumor cells with higher metastatic protential bind better to laminin than to fibronectin.[103] Finally, antibodies to laminin,[103,122] its minimal cell attachment peptide (YIGSR)[123] and protease-derived laminin fragments[45] are capable of binding to the laminin receptor, block the binding of malignant cells to laminin and show antiinvasive and antimetastatic activity in an experimental model of metastasis.

Thus, cancer cell adhesion to the basement membranes (laminin) and extracellular matrix (fibronectin) plays a critical role in cancer invasion. Antimetastatic action in vivo of synthetic peptide cell adhesion epitopes derived both from fibronectin and laminin strongly support this concept.[16,123-125] Glycodeterminants of cell adhesion may play an important role in attachment of cancer cells to the basement membranes and extracellular matrix since both laminin and fibronectin contain cell adhesion glycoepitopes, and participation of specific carbohydrate structures

in cancer cell recognition of laminin and fibronectin has been shown experimentally (see above).

Chemotaxis is cellular locomotion that is directed in response to a concentration gradient of a chemical mediator in solution.[126] If cells migrate in the direction of increasing concentration of mediator, chemotaxis is positive, and if migration is toward lower concentration chemotaxis is negative. Cancer cells are able to release into extracellular medium a number of glycomacromolecules, e.g., glycolipids, glycoproteins, glycoamines, etc., that contain glycodeterminants of cell adhesion in biologically active form. Therefore, cell adhesion glycoepitope-mediated negative chemotaxis may be one of the key factor determining direction of cancer cell migration from the zone of primary tumor.

In adult organisms the tissue cells respect the tissue boundaries and refrain from wandering from the confines of the parent tissue into adjacent tissue. In this context the organization of tissues and organs is stable after completion of cell and tissue movement during embryonic morphogenesis. However, most cell types are capable of active locomotion and latent locomotion is revealed during wound healing[127-132] in cells transferred to cell culture[90] and during physiological regeneration of epithelia. It has been suggested that a variety of processes may contribute to stabilizing the organization of tissues composed of potentially motile cells.[38,70,71]

The cells of coherent tissues are normally stationary as a consequence of a various static processes such as: a) contact inhibition of pseudopodial activity (pseudopodial protrusion and cellular locomotion is suppressed at areas of cell-cell contact); b) impenetrable interstitial matrix prevents any translocation of cells; and c) strong cell-cell adhesion that is mediated by cell junctions (desmosomes, gap and tight junctions) or by intercellular matrix proteins (fibronectin, collagen, laminin) prevents the break of cell-cell contacts by the force of locomotion. The dynamic processes, which do not prevent movement of cells within the parent tissues but prevent the movement across the tissue boundaries, consist of mechanical barriers at tissue borders such as basal lamella, and the adhesive affinity of homotypic tissue that favors homotypic tissue recognition.[38] The invasion mechanisms are related to an ineffectual display or subversion of the physiological processes that contribute to the tissue and organ stability.[38] The morphogenetic movements of cells and tissues during enbryogenesis are invasive in character and invasive behavior is displayed by motile blood cells,[93,133] cells involved in wound healing[127-130] and cancer cells.[134,135] Most of the processes that may contribute to the stabilization of tissue organization are dependent on the ability of cells to develop cell type-specific secondary strong and stable adhesion. Therefore, the ability of cancer cells to display the initial primary carbohydrate-mediated step of cellular recognition and adhesion and impairment of secondary strong attachment, intracellular signaling and secondary adhesion-dependent cellular function are the key factors determining abnormal social behavior of malignant cells, e.g., loss of stable organization of parent tissue, invasive growth and metastatic dissemination. The phenomenon of in vitro cell sorting which shows the dynamic character of tissue stability, supports this conclusion. There are two stages of cell sorting when cell aggregates are produced from suspensions of dissociated cells containing two or more cell types.[82,136] The initial aggregates are disordered collections of different cell types. However, when maintained in culture for one or more days, the disordered arrangement is replaced as the cells sort out into homogeneous tissue domains. Numerous studies indicate that cancer cells preserve in vitro the tissue affinity function[62,137-139] and are able to form multicell spheroids similar to the malignant tumors in vivo biochemical, immunocytochemical, morphological and cytological characteristics.

REFERENCES

1. Hedman, K., Kurkinen, M., Alitalo, K., Vaheri, A., Johansson, S., Hook, M. (1979) Isolation of the pericellular matrix of human fibroblast cultures. J. Cell Biol., 81:83-91.

2. Kleinman, H.K., Klebe, R.J., Martin, G.R. (1981) Role of collagenous matrices in the adhesion and growth of cells. J. Cell Biol., 88:473-485.

3. Alitalo, K., Vaheri, A. (1982) Pericellular matrix and malignant transformation. Adv. Cancer Res., 37:111-158.

4. Liotta, L.A. (1986) Tumor invasion and metastases—role of the extracellular matrix. Cancer Res., 46:1-7.

5. Quigley, J.P. (1976) Association of a protease (plasminogen activator) with a specific membrane fraction isolated from transformed cells. J. Cell Biol., 71:472-486.

6. Chen, W.-T., Olden, K., Bernard, B.A., Chu, F.-F. (1984) Expression of transformation-associated protease(s) that degrade fibronectin at cell contact sites. J. Cell Biol., 98:1546-1555.

7. Tryggvason, K., Hoyhtya, M., Salo, T. (1987) Proteolytic degradation of extracellular matrix in tumor invasion. Biochim. Biophys. Acta., 907:191-217.

8. Quigley, J.P., Goldfarb, R.H., Scheiner, C., O'Donnel-Tormey, J., Yeo, T.K. (1980) Plasminogen activator and the membrane of transformed cells. Prog. Clin. Biol. Res., 41:773-795.

9. Bronner-Fraser, M. (1986) An antibody to a receptor for fibronectin and laminin perturbs cranial neural crest development in vivo. Dev. Biol., 117:528-536.

10. Chen, W.-T., Chen, J.M., Mueller, S.C. (1986b) Coupled expression and colocalization of 140K cell adhesive molecules, fibronectin and laminin during morphogenesis and cytodifferentiation of chick lung cells. J. Cell Biol., 103:1073-1090.

11. Duband, J.-L., Rocher, S., Chen, W.-T., Yamada, K.M., Thiery, J.P. (1986) Cell adhesion and migration in the early vertebrate embryo: location and possible role of the putative fibronectin receptor. J. Cell Biol., 102:160-178.

12. Duband, J.-L., Dufour, S., Yamada, K.M., Thiery, J.P. (1988) The migratory behavior of avian embryonic cells does not require phosphorylation of the fibronectin receptor complex. FEBS (Fed. Eur. Biochem. Soc.) Lett., 230:181-185.

13. McClay, D.R., Ettensohn, C.A. (1987) Cell adhesion in morphogenesis. Annu. Rev. Cell Biol., 3:319-345.

14. Boucaut, J.C., Darribere, T., Poole, T.J., Aoyama, H., Yamada, K.M., Thiery, J.P. (1984) Biologically active synthetic peptides as probes of embryonic development: a competitive peptide inhibitor of fibronectin function inhibits gastrulation in amphibian embryos and neural crest cell migration in avain embryos. J. Cell Biol., 99:1822-1830.

15. Humphries, M.J., Akiyama, S.K., Komoriya, A., Olden, K., Yamada, K.M. (1986a) Identification of an alternatively spliced site in human plasma fibronectin that mediates cell type-specific adhesion. J. Cell Biol., 103:2637-2647.

16. Humphries, M.J., Olden, K., Yamada, K.M. (1986b) A synthetic peptide from fibronectin inhibits experimental metastasis of murine melanoma cells. Science (Wash DC), 233:467-470.

17. Singer, I.I., Scott, S., Kawka, D.W., Kazazis, D.M., Gailit, J., Ruoslahti, E. (1988) Cell surface distribution of fibronectin and vitronectin receptors depends

on substrate composition and extracellular matrix accumulation. J. Cell Biol., 106:2171-2182.

18. Roman, J., LaChance, R.M., Broekelmann, T.J., Kennedy, C.J.R., Wagner, E.A., Carter, W.G., McDonald, J.A. (1989) The fibronectin receptor is organized by extracellular matrix fibronectin: implication for oncogenic transformation and for cell recognition of fibronectin matrices. J. Cell. Biol., 108:2529-2543.

19. Chen, W.-T., Wang, J., Hasegawa, T., Yamada, S.S., Yacroda, K.M. (1986a) Regulation of fibronectin receptor distribution by transformation, exogeneous fibronectin and synthetic peptides. J. Cell Biol., 103:1649-1661.

20. Wagner, D.D., Ivatt, R., Destree, A.T., Hynes, R.O. (1981) Similarities and differences between the fibronectins of normal and transformed hamster cells. J. Biol. Chem., 256:11708-11715.

21. Yamada, K.M. (1978) Immunological characterization of a major transformation-sensitive fibroblast cell surface glycoprotein. Localization, redistribution and role in cell shape. J. Cell Biol., 78:520-541.

22. Fegan, J.B., Sobel, M.E., Yamada, K.M., de Crombrugghe, B., Pastan, I. (1981) Effects of transformation on fibronectin gene expression using cloned fibronectin cDNA. J. Biol. Chem., 256:520-525.

23. Chen, J.-M., Chen, W.-T. (1987) Fibronectin-degrading proteases from the membranes of transformed cells. Cell., 48:193-203.

24. Hirst, R., Horwitz, A., Buck, C., Rohrschneider, L. (1986) Phosphorylation of the fibronectin receptor in cells transformed by oncogenes that encode tyrosine kinases. Proc. Natl. Acad. Sci USA., 83:6470-6474.

25. McDonald, J.A., Quade, B.J., Broekelmann, T.J., La Chance, R., Forseman, K., Hasegawa, K.E., Akiyama, S. (1987) Fibronectins cell-adhesive domain and an amino terminal matrix assembly domain participate in its assembly into fibroblast pericellular matrix. J. Biol. Chem., 272:2957-2967.

26. Menko, A.S., Boettiger, D. (1987) Occupation of the extracellular matrix receptor, integrin, is a control point for myogenic differentiation. Cell., 51:51-57.

27. Patel, V.P., Lodish, H.F. (1988) A fibronectin matrix is required for differentiation of murine erythroleukemia cells into reticulocytes. J. Cell Biol., 105:3105-3118.

28. Bitterman, P.B., Rennard, S.I., Adelberg, S., Crystal, R.G. (1983) Role of fibronectin as a growth factor for fibroblasts. J. Cell Biol., 97:1925-1932.

29. Holderbaum, D., Ehrhart, L.A. (1986) Substratum influence on collagen and fibronectin biosynthesis by arterial smooth muscle cells in vitro. J. Cell. Physiol., 126:216-224.

30. Yamada, K.M. (1983) Cell surface interactions with extracellular materials. Annu. Rev. Biochem., 52:761-799.

31. Mosher, D.F. (1984) Physiology of fibronectin. Annu. Rev. Med., 35:561-575.

32. Matsuura, H., Hakomori, S. (1985) The oncofetal domain of fibronectin defined by monoclonal anti-body FDC-6: its presence in fibronectins from fetal and tumor tissues and its absence in those from normal adult tissues and plasma. Proc. Natl. Acad. Sci USA., 82:6517-6521.

33. Matsuura, H., Takio, K., Titani, K., Greene, T., Levery, S.B., Salyan, M.E.K., Hakomori, S. (1988) The oncofetal structure of human fibronectin defined by monoclonal antibody FDC-6. J. Biol. Chem., 263:3314-3322.

34. Nichols, E.J., Fenderson, B.A., Carter, W.G., Hakomori, S. (1986) Domain-specific distribution of carbohydrates in human fibronectins and the transformation-dependent translocation of branched type 2 chain defined by monoclonal antibody C6. J. Biol. Chem., 261:11295-11301.

35. Lisowska, E., Wasniowaka, K. (1986) Eur. J. Biochem., 88:247-252.

36. Sadler, J.E., Paulson, J.C., Hill, R.L. (1979) The role of sialic acid in the expression of human MN blood group antigens. J. Biol. Chem., 254:2112-2119.

37. LeDouarin, N.M. (1984) Cell migration in embryos. Cell., 38:353-360.

38. Armstrong, P.B. (1985) The control of cell motility during embryogenesis. Cancer and Metastasis Rev., 4:59-80.

39. Thiery, J.P., Duband, J.L., Tucker, G.C. (1985) Annu. Rev. Cell Biol., 1:91-113.

40. Krotoski, D.M., Domingo, C., Bronner-Fraser, M. (1986) J. Cell Biol., 103:1061-1071.

41. Cheresh, D.A., Harper, J.R., Schulz, G., Reisfeld, R.A. (1984). Localization of the gangliosides G_{D2} and G_{D3} in adhesion plaques and on the surface of human melanoma cells. Proc. Natl. Acad. Sci., 81:5767-5771.

42. Cheresh, D.A., Klier, F.G. (1986). Disialoganglioside G_{D2} distributes preferentially into substrate-associated microprocesses in human melanoma cells during their attachment to fibronectin. J. Cell Biol., 102:1887-1897.

43. Runyan, R.B., Maxwell, G.D., Shur, B.D. (1986) Evidence for a novel enzymatic mechanism of neural crest cell migration on extracellular glycoconjugate matrices. J. Cell Biol., 102:432-441.

44. Rao, C.N., Barsky, S.H., Terranova, V.P., Liotta, L.A. (1983) Isolation of a tumor cell laminin receptor. Biochem. Biophys. Res. Commun., 111:804-808.

45. Barsky, S.H., Rao, C.N., Williams, J.E., Liotta, L.A. (1984) Laminin molecular domains which alter metastasis in a murine model. J. Clin. Invest., 74:843-848.

46. Runyan, R.B., Versalovic, J., Shur, B.D. (1988) Functionally distinct laminin receptors mediate cell adhesion and spreading: the requirement for surface galactosyltransferase in cell spreading. J. Cell Biol., 107:1863-1871.

47. Graf, J., Iwamoto, Y., Sasaki, M, Martin, G.R., Kleinman, H.K., Robey, F.A., Yamada, Y. (1987) Identification of an amino acid sequence in laminin mediating cell attachment, chemotaxis and receptor binding. Cell, 48:989-996.

48. Eckstein, D.J., Shur, B.D. (1989) Laminin induces the stable expression of surface galactosyltransferase on lamellipodia of migrating cells. J. Cell Biol., 108:2507-2517.

49. Eckstein, D.J., Shur, B.D. (1988) Cell surface galactosyltransferase interaction with laminin during cell migration on basal lamina. In: Altered Glycosylation in Tumor Cells. Editors: Ch.L. Reading, S.-I. Hakomori, D.M. Markus, N.Y., Alan R. Liss, pp. 217-229.

50. Loewenstein, W.R. (1979) Junctional intercellular communication and the control of growth. Biochim. Biophys. Acta, 560:1-65.

51. Lorand, L., Losowsky, M.S., Miloszewski, K.J.M. (1980) Human factor XIII: fibrin-stabilizing factor. Prog. Hemostasis Thromb., 5:245.

52. McDonagh, J. (1987) In: Hemostasis and Thrombosis. (R.W. Colman, W. Hirsh, V.J. Marder and E.W. Salzman, eds.), J.B. Lippincott Co., Philadelphia, p.289.

53. Folk, J.E. (1980) Transglutaminases. Ann. Rev. Biochem., 49:517.

54. Lorand, L., Conrad, S.M. (1984) Transglutaminases. Molec. Cell. Biochem., 58:9.

55. Mosher, D.F. (1987) In: Hemostasis and Thrombosis. (R.W. Colman, J. Hirsh, V.J. Marder and E.W. Salzman, eds.), J.B. Lippincott Co., Philadelphia, p. 210.

56. Mosher, D.F. (1975) J. Biol. Chem., 250:6614.

57. Mosher, D.F., Shad, P.E., Kleiman, H.K. (1979) J. Clin. Invest., 64:781.

58. Mosher, D.F., Johnson, R.B. (1985) J. Biol. Chem., 258:6595.

59. Williams, F.C., Janmey, P.A., Johnson, R.B., Mosher, D.F. (1983) J. Biol. Chem., 258:5911.

60. Mosher, D.F. (1976) J. Biol. Chem., 251:1639.

61. Glinsky, G.V. (1989) Glycoamines: Biochemistry of a new class of humoral tumor markers. J. Tumor Marker Oncology, 4:193-221.

62. Glinsky, G.V. (1992a) Glycoamines: Structural-functional characterization of a new class of human tumor markers. In: Serological Cancer Markers. Editor: S. Sell. The Humana Press, Totowa, N.J., Chapter 11, pp. 233-260.

63. Glinsky, G.V., Surgova, T.M., Sidorenko, M.V., Vinnitsky, V.B., Kolesnik, L.A., Shvachko, L.A. (1990) Interaction of glycoamines and polyamines within the system of α2 macroglobulin-dependent regulation of plasma transglutaminase. A possible molecular mechanism of invasion and implantation in embryogenesis and cancer. J. Tumor Marker Oncology, 5:161-166.

64. Brownlee, M., Vlassara, H., Cerami, A. (1984) Non-enzymatic glycosylation and the pathogenesis of diabetic complications. Ann. Intern. Med., 101:527.

65. Ganea, E. (1987) Rev. Roum. Biochim., 24:217.

66. Tarsio, J.F., Reger, L.A., Furcht, L.T. (1987) Decreased interaction of fibronectin, type IV collagen, and heparin due to nonenzymatic glycation. Implications for diabetes mellitus. Biochemistry, 26:1014.

67. Katz, M.J., Lasek, R.J. (1980) Guidance cue patterns and cell migration in multicellular organisms. Cell Motil., 1:141-157.

68. Trinkaus, J.P. (1976) On the mechanism of metazoan cell movements. In: The cell surface in animal embryogenesis and development. Editors: G. Poste and G.L. Nicholson, North Holland, New York, pp. 225-329.

69. Trinkaus, J.P. (1984) Cells into organs. (2nd ed.). Prentice-Hall, Inc., Englewood Cliffs, N.J.

70. Armstrong, P.B. (1977) Cellular positional stability and intercellular invasion. BioScience, 27:803-809.

71. Armstrong, P.B. (1984) Invasiveness of non-malignant cells. In: Invasion: Experimental and clinical implications. Editors: M.M. Mareel and Calman, K.C., Oxford University Press, Oxford, pp. 126-167.

72. Tosney, K.W. (1982) The segregation and early migration of cranial neural crest cells in the avian embryo. Dev. Biol., 89:13-24.

73. Newgreen, D., Gibbins, I. (1982) Factors controlling the time of onset of the migration of neural crest cells in the fowl embryo. Cell Tissue Res., 224:145-160.

74. Nicholas, D.H. (1981) Neural crest formation in the head of the mouse embryo as observed using a new histological technique. J. Embryol. Exp. Morph., 64:105-120.

75. Carter, S.B. (1967) Haptotaxis and the mechanism of cell motility. Nature, 213:256-160.

76. Harris, A. (1973) Behavior of cultured cells on substrata of variable adhesiveness. Exp. Cell Res., 77:285-297.

77. Letourneau, P.C. (1975) Possible roles for cell-to-substratum adhesion in neuronal morphogenesis. Dev. Biol., 44:77-91.

78. Letourneau, P.C. (1975) Cell-to-substratum adhesion and guidance of axonal elongation. Dev. Biol., 44:92-101.

79. Gustafson, T., Wolpert, L. (1961) Studies on the cellular basis of morphogenesis in the sea urchin embryo. Directed movements of primary mesenchyme cells in normal and vegetalized larvae. Exp. Cell Res., 24:64-79.

80. Holtfreter, J. (1939) Tissue affinity, a means of embryonic morphogenesis. Arch. Exp. Zellforsch., 23:169-209. [English translation in Willier, B.H., Oppenheimer, J.M. (eds) Foundations of experimental embryology. Prentice-Hall, Inc., Englewood Cliffs, N.J., pp.186-225].

81. Holtfreter, J. (1944) Experimental studies on the development of the pronephros. Rev. Canad. Biol., 3:220-250.

82. Townes, P., Holtfreter, J. (1955) Directed movements and selective adhesion of embryonic amphibian cells. J. Exp. Zool., 128:53-120.

83. Steinberg, M.S. (1963) Reconstitution of tissues by dissociated cells. Science, 141:401-408.

84. Steinberg, M.S. (1970) Does differential adhesion govern self-assembly processes in histogenesis? Equilibrium configurations and the emergence of a hierarchy among populations of embryonic cells. J. Exp. Zool., 173:395-434.

85. Steinberg, M.S. (1978) Specific cell ligands and the differential adhesion hypothesis: How do they fit together? In: Specificity of embryological interactions. Receptors and Recognition, Series B, Vol. 4, Editor: D.R. Garrod, Chapman and Hall, London, pp. 99-130.

86. Steinberg, M.S. (1978) Cell-cell recognition in multicellular assembly: Levels of specificity. In: Cell-cell recognition; Symp. Soc. Exp. Biol. (32). Editor: A.S. G. Gurtis, Cambridge University Press, Cambridge, pp. 25-49.

87. Phillips, H.M., Steinberg, M.S. (1969) Equilibrium measurements of embryonic chick cell adhesiveness I. Shape equillibrium in centrifugal fields. Proc. Natl. Acad.Sci., 64:121-127.

88. Phillips, H.M., Davis, G.S. (1978) Liquid-tissue mechanics in amphibian gastrulation: Germ-layer assembly in Rana pipiens. Amer. Zool., 18:81-93.

89. Roos, E. (1983). Cellular adhesion, invasion and metastasis. Biochim. Biophys. Acta., 738:263-284.

90. Abercrombie, M. (1970) Contact inhibition in tissue culture. In Vitro, 6:128-142.

91. Abercrombie, M., Ambrose, E.J. (1958) Interference microscope studies of cell contacts in tissue culture. Exp. Cell Res., 15:332-345.

92. Heaysman, J.E.M. (1978) Contact inhibition of locomotion: a reappraisal. Int. Rev. Cytol., 55:49-66.

93. Armstrong, P.B., Lackie, J.M. (1975) Studies on intercellular invasion *in vitro* using rabbit peritoneal neutrophil granulocytes (PMNS) I. Role of contact inhibition of locomotion. J. Cell Biol., 65:439-462.

94. Martz, E., Steinberg, M.S. (1972) Contact inhibition of what? An analytical review. J. Cell Physiol., 81:25-38.

95. Weiss, P. (1958) Cell contact. Int. Rev. Cytol., 7:391-423.

96. Vracko, R. (1974) Basal lamina scaffold—Anatomy and significance for maintenance of orderly tissue structure. Am. J. Pathol., 77:314-346.

97. Gross, J.L., Moscatelli, D., Jaffe, E.A., Rifkin, D.B. (1982) Plasminogen activator and collagenase production by cultured capillary endothelial cells. J. Cell Biol., 95:974-981.

98. Valinsky, J.E., Reich, E., Le Douarin, N. (1981) Plasminogen activator in the bursa of Fabricius: Correlations with morphogenetic remodeling and cell migrations. Cell, 25:471-476.

99. Jones, P.A., De Clerck, Y.A. (1982) Extracellular matrix destruction by invasive tumor cells. Cancer Metastasis Rev., 1:289-317.

100. Liotta, L.A., Thorgeirsson, U.P., Garbisa, S. (1982) Role of collagenases in tumor cell invasion. Cancer Metastasis Rev., 1:277-288.

101. Quigley, J.P. (1979) Proteolytic enzymes of normal and malignant cells. In: Surfaces of normal and malignant cells. Editor: Hynes, R.D., Wiley, Chichester, pp. 247-285.

102. Timpl, R., Dziadek, M. (1986) Structure, development and molecular pathology of basement membrane. Int. Rev. Exp. Pathol., 29:1-112.

103. Terranova, V.P., Liotta, L.A., Russo, R.G., Martin, G. R. (1982) Role of laminin in the attachment and metastasis of murine tumor cells. Cancer Res., 42:2265-2269.

104. Liotta, L.A. (1984) Tumor invasion and metastasis: role of the basement membrane. Am. J. Pathol., 117:335-348.

105. Terranova, V.P., Hujanen, E.S., Martin, G.R. (1986) Basement membrane and the invasive activity of metastatic tumor cells. J. Natl. Cancer Inst., 77:311-316.

106. Nakajima, M., Welch, D.R., Irimura, T., Nicolson, G.L. (1986) Basement membrane degradative enzymes as possible markers of tumor metastasis. In: Cancer Metastasis: Experimental and Clincial Strategies. Editors: Welch, D.R., Bhuyan, B.K., Liotta, L.A. Alan R. Liss, N.Y., p. 113-122.

107. Turpeenniemi-Hujanen, T., Thorgeirsson, U.P., Rao, C.N., Liotta, L.A. (1986) Laminin increases the release of type IV collagenase from malignant cells. J. Biol.Chem., 261:1883-1889.

108. Reich, R., Stratford, B., Klein, K., Martin, G.R., Mueller, R.A., Fuller, G.C. (1988) Inhibitors of collagenase IV and cell adhesion reduce the invasive activity of malignant tumor cells. In: 1988 Metastasis. Editors: Wiley, Chichester., Ciba Foundation Simposium 141., pp. 193-210.

109. Thompson, E.W., Reich, R., Martin, G.R., Albini, A. (1988) Factors regulation basement membrane invasion by tumor cells. In: Breast Cancer. Cellular and Molecular Biology. Editors: Lippman, M.E., Dickson, R.B. Kluwer Academic Publishers. Boston/Dordrecht/London., pp. 239-249.

110. Laiho, M., Keski-Oja, J. (1989) Growth factors in the regulation of pericellular proteolysis: A review. Cancer Res., 49:2533-2553.

111. Stopelli, M.P., Corti, A., Soffientini, A., Cassani, G., Blasi, F., Assoian, R.K. (1985) Differentiation-enhanced binding of the amino-terminal fragment of human urokinase plasminogen activator to a specific receptor on U 937 monocytes. Proc. Natl. Acad. Sci. USA, 82:4939-4943.

112. Mullins, D.E., Rorlich, S.T. (1983) The role of proteinases in cellular invasiveness. Biochem. Biophys. Acta., 695:177-214.

113. Ossowski, L. (1988) Plasminogen activator dependent pathways in the dissemination of human tumor cells in the chick embryo. Cell, 52:321-328.

114. Mignotti, P., Tsubai, R., Robbins, E., Rifkin, D.B. (1989) In vitro angiogenesis of the human amniotic membrane: requirement for basic fibroblast growth factor-induced proteinases. J. Cell. Biol., 108:671-682.

115. Axelrod, J.H., Reich, R., Miskin, R. (1989) Expression of human recombinant plasminogen activators enhances invasion and experimental metastasis of H-ras-transformed NIH 3T3 cells., Mol. Cell Biol., 9:2133-2141.

116. Laiho, M., Saksela, O.,reasen, P.A., Keski-Oja, J. (1986) Enhanced production and extracellular deposition of the endothelial-type plasminogen activator inhibitor in cultured human lung fibroblasts by

transforming growth factor-β. J. Cell Biol., 103:2403-2410

117. Lee, L.-S., Weinstein, I.B. (1978) Epidermal growth factor, like phorbol esters, induces plasminogen activator in HeLa cells. Nature (London), 274:696-697.

118. Matrisian, L.M., Leroy, P., Ruhlmann, C., Gesnel, M.-C., Breathnach, R. (1986) Isolation of the oncogene and epidermal growth factor-induced transin gene: complex control in rat fibroblasts. Mol. Cell Biol., 6:1679-1686.

119. Kleinman, H.K., Cannon, F.B., Laurie, G.W., Hassell, J.R., Aumailley, M., Terranova, V.P., Martin, G.R., Dubois-Dalcq, M. (1985) Biological activities of laminin. J. Cell Biochem., 27:317-325.

120. Wewer, U.M., Liotta, L.A., Jaye, M., Ricca, G.A., Drohan, W.N., Claysmith, A.P., Rao, C.N., Wirth, P., Coligan, J.E., Albrechtsen, R., Mudryj, M., Sobel, M.E. (1986) Altered levels of laminin receptor mRNA in various human carcinoma cells that have different abilities to bind laminin. Proc. Natl. Acad. Sci. USA, 83:7137-7141.

121. Terranova, V.P., Rao, C.N., Kalebic, T., Margulies, I.M., Liotta, L.A. (1983) Laminin receptor on human breast carcinoma cells. Proc. Natl. Acad. Sci. USA, 80:444-448.

122. Terranova, V.P., Williams, J.E., Liotta, L.A., Martin, G.R. (1984) Modulation of the metastatic activity of melanoma cells by laminin and fibronectin. Science, 226:982-985.

123. Iwamoto, Y., Robey, F.A., Graf, J., Sasaki, M., Kleinman, H,K., Yamada, Y and Martin, G.R. (1987) YIGSR, a synthetic laminin pentapeptide inhibits experimental metastasis formation. Science, 238:1132-1134.

124. Saiki, I., Iida, J., Murata, J., Ogawa, R., Nishi, N., Sugimura, K., Tokura, S., Azuma, I. (1989) Inhibition of the metastasis of murine malignant melanoma by synthetic polymeric peptides containing core sequences of cell-adhesive molecules. Cancer Res., 49:3815-3822.

125. Ugen, K.E., Mahalingam, M., Klein, P.A., Kao, K.-J. (1988) Inhibition of tumor cell-induced platelet aggregation and the experimental tumor metastasis by the synthetic Gly-Arg-Gly-Asp-Ser peptide. J.

Natl. Cancer Inst., 80:1461-1466.

126. Harris, H. (1954) Role of chemotaxis in inflamation. Physiol. Rev., 34:529.

127. Cliff, W.J. (1963) Observations on healing tissue: A combined light and electron microscopic investigation. Phil. Trans. Roy. Soc. London B, 246:305-325.

128. Cliff, W.J. (1965) Kinetics of wound healing in rabbit ear chambers, a time lapse cinemicroscopic study. Quant J. Exp. Physiol., 50:79-89.

129. Schoefl, G.I. (1964) Electron microscopic observations on the regeneration of blood vessels after injury. Ann. NY Acad. Sci., 116:789-802.

130. Jennings, M.A., Florey, H.W. (1970) Healing. In: General pathology 4th edn. Editor: Florey, H. W., W.B. Saunders, Philadelphia, pp. 480-548.

131. Lash, J. (1955) Studies on wound closure in urodeles. J. Exp. Zool., 128:13-28.

132. Radice, G.P. (1980) The spreading of epithelial cells during wound closure in Xenopus larvae. Dev. Biol., 76:26-46.

133. Armstrong, P.B. (1980) Invasiveness of neutrophil leukocytes. In: Cell movement and neoplasia. Editor: De Brabander, M., Pergamon Press, Oxford, pp. 131-147.

134. Strauli, P., Weiss, L. (1977) Cell locomotion and tumor penetration. Europ. J. Cancer, 13:1-12.

135. Mareel, M.M. (1980) Recent aspects of tumor invasiveness. Int. Rev. Exp. Pathol., 22:65-129.

136. Armstrong, P.B. (1971) Light and electron microscope studies of cell sorting in combinations of chick embryo neural retina and retinal pigment epithelium. Wilhelm Roux Arch, 168:125-141.

137. Glinsky, G.V. (1992b) The blood group antigens (BGA)-related glycoepitope. A key structural determinants in immunogenesis and cancer pathogenesis. Critical Reviews in Oncology/Hematology, 12:151-166.

138. Mueller-Kleiser, W. (1987) Multicellular spheroids. A review on cellular aggregates in cancer research. J. Cancer Res. Clin. Oncol., 113:101-122.

139. Sutherland, R.M. (1988) Cell and environment interactions in tumor microregions: The multicell spheriod model. Science, 240:177-184.

CHAPTER 4

BGA-RELATED GLYCOEPITOPES AS KEY STRUCTURAL DETERMINANTS IN IMMUNOGENESIS

One of the key features of the cells of the immune system is the rapid transition between adherent and nonadherent states. Three families of adhesion receptors of the immune system have been discovered: the immunoglobulin superfamily; the integrin family; and the selectins. They are involved in adhesive interactions that direct lymphocyte localization and migration, determine lymphocyte homing to different lymphoid organs and direct neutrophil localization in inflammation.[1] Recently, dramatic progress has been made in our understanding of the molecular mechanism of leukocyte-endothelial cell interaction and the role of specific carbohydrate determinants in these processes. Here we will summarize current data concerning the mechanisms that determine the site specificity of leukocyte-endothelium interaction and the role in the endothelial cell-leukocyte recognition of the BGA-related glycoepitoipes. The mechanism will be considered that may be responsible for selective expression of certain BGA-related glycodeterminants on specific leukocyte subsets.

THE SITE SPECIFICITY OF LEUKOCYTE—ENDOTHELIAL CELL RECOGNITION, ADHESION AND EXTRAVASATION

Rapid transition between adherent and nonadherent states (and vice versa) is one of key biological features of the immune cells. To patrol body homeostasis effectively, the cells of the immune system circulate as nonadherent cells in the blood and lymph and migrate as adherent cells through tissues. The immune cells are able to cross endothelial and basement membrane at specific sites to aggregate at places of infection and/or adhere to cells bearing foreign antigen. Thus patrolling the body, the lymphocytes leave the blood, migrate through lymphoid organs and other tissues, enter the lymphatics and return to the blood through the thoracic duct. It is important that priming by a specific antigen may alter the surface phenotype, enabling selective recirculation of lymphocytes to the particular type of secondary lymphoid organs (lymph nodes) where the specific antigen was first encountered.[2,3] Lymphocytes in the blood enter lymph nodes by binding to specialized "high" endothelial cells and "recirculation" or "homing" receptors on lymphocytes have been defined by monoclonal antibodies that block binding of lymphocytes to the high endothelial cells of specific types of lymph nodes.[2-4]

Specific binding and immigration of lymphocytes to the high endothelial venules in lymph nodes is a sequential cooperative multistep "cascade" process involving several receptors on the lymphocytes and counter-receptors on the endothelium.[1,5]

The recruitment of leukocytes from the blood into the tissues—leukocyte extravasation—is regulated by selective mechanisms of leukocyte-endothelial cell recognition. The physiologic trafficking of lymphocytes and cellular response to tissue damage and inflammation show extraordinary specificity in relation to the tissue site or organ involved.[5] The tissue-specific selective interaction of lymphocyte subsets with high endothelial venules in lymph nodes and specific recruitment of neutrophils in inflammation are the best examples. However, this specific leukocyte-endothelial cell recognition cannot be explained by a lock-and-key model, since individual receptors participate in multiple leukocyte-endothelial cell interactions that are quite independently regulated in vivo.[5] A general model of leukocyte-endothelial cell recognition has been proposed in which this process requires at least three sequential events.[5] Interaction is initiated by transient and reversible adhesion through binding of constitutively functional leukocyte adhesion receptors to endothelial cell counterreceptors. Typically, this step requires endothelial cell activation and such adhesion is mediated by lectin-carbohydrate interactions involving leukocyte or endothelial cell selectins and their oligosaccharide ligands.[6-11] The second event is activation of the leukocyte by a specific chemoattractant or cell contact which activates the secondary adhesion receptors (the integrin or immunologbulin superfamily member). The function of those adhesion receptors is activation-dependent and their interaction with endothelial counterreceptors results in strong, stable attachments, completing the recognition process.[5]

The initial reversible adhesion, activation and activation-dependent stable binding may be the common mechanism to many or all specific leukocyte-endothelial cell recognition reactions. The specificity of those interactions is determined by unique combinations of initial adhesion receptor-ligand and activating receptor-ligand pairs, the number and combination of which is the product of several sequential recognition events, each representing a go/no go decision point.[5] A given highly effective receptor-ligand pair participates in several recognition events and this molecular conservatism does not sacrifice specificity of leukocyte-endothelial cell recognition and leukocyte extravasation.

BGA-Related Glycoepitopes as Key Structural Determinants of Leukocyte Recognition and Adhesion: Implication for Critical Biological Functions of the Immune Cells

Three families of immune system adhesion receptors have been discovered: the immunoglobulin superfamily, the integrin family and the selectins. They are involved in adhesive interactions that direct lymphocyte localization and migration, determine lymphocyte homing to different lymphoid organs and neutrophil localization in inflammation.[1,4,5] Carbohydrate ligands for selectins have recently been discovered: Sialosyl-LeX, Sialosyl-LeA and LeX have been identified as recognition structures for ELAM-1 and CD62.[6-9,11] Neutrophils bear LeX both on glycolipids and at the termini of N- and O-linked oligosaccharides[12,13] Finally, the study of how leukocytes roll on a selectin at physiologic flow rates elegantly shows that rolling on a selectin is a prerequisite for activiation-induced adhesion through integrins.[10] Thus, these results clearly demonstrate the role of BGA-related glycodeterminants as specific mediators of the early initial step of leukocyte adhesion.

Based on these and other facts, we have suggested[14-19] that the processes of thymic education and antigen presentation are accompanied by certain changes in the glycosylation pattern of the cellular membranes. These changes include the presentation of the BGA-related glycoepitopes—in particular, of their cryptic forms. Aberrant glycosylation may occur as a result of binding of foreign peptides to MHC and subsequent changes in glycosylation patterns of peptide-MHC complexes and/or as a result of presentation of peptide-MHC complexes on the cell membrane and exposition of new, previously masked, glycodeterminants. Numerous experimental data are now available that confirm this suggestion for antigen presentation.

Evidence has suggested that T cell recognition of stimulator (antigen-presenting) cells may be influenced by the state of glycosylation of the stimulator cells.[20-25] Recognition of specific carbohydrate determinants, probably associated with HLA antigens on the surface of antigen presenting cells, may play a key regulatory role in function of both MHC class 1 and MHC class II restricted T cells. Black et al[25] have suggested a regulatory role for asparagine-linked oligosaccharide in the H-2 restricted cytolysis of virus-infected cells. Tunicamycin treatment of P 815 cells before and during infection with vesicular stomatitis virus (VSV) inhibited by about 50% the lysis of infected P 815 cells by VSV-immune H-2 identical killer cells.[25] Furthermore, a numerous data indicate the role of oligosaccharides in cellular recognition by thymic lymphocytes:[20] a) selective removal of cell surface sialic acids from the surface of antigen-presenting cells supports their recognition by MHC class II reactive T cells;[23,24] b) murine B cell and spleen adherent cell Ia antigens display a different glycosylation pattern;[21] c) a potential regulatory function has been proposed for carbohydrate side chains on Ia molecules in T cells with discrimination between Ia antigens expressed on allogenic accessory cells and B cells.[22] Sialic acids on cell surface molecules, inlcuding MHC, may play a role in antigen presentation since removal of sialic acids, by neuraminidase, can restore specific responses to nonresponder antigen-presenting cells and effective antigen presentation occurs in cells expressing an increased number of histocompatibility antigens with decreased sialylation levels.[26] Cells infected with AIDS virus or cytomegalovirus start to express high levels of LeY and LeX glycoepitopes, respectively.[27,28] This relatively nonspecific mechanism of the presentation of BGA-related glycoepitopes may be involved in the process of homotypic sorting of cells at subsequent stages of formation of immune response. In view of this, natural anticarbohydrate antibodies, identified in blood serum from most healthy individuals,[29] may fulfill a double function: to accomplish the complement-dependent lysis of cells containing high-density BGA-related glycoepitopes, e.g., in thymic education it may lead to negative selection of T-cells whose major histocompatibility complexes have a high affinity for their own presented antigens, and to favor the homotypic selection of cells revealing a medium or low density level of the BGA-related glycoepitopes.

The recent study of Huesmann et al[30] shows that the binding of TCR to thymic MHC molecules rescues CD4+8+ cells from programmed cell death and induces first up regulation of the TCR level and then differentiation into CD4+8- or CD4 -8+ cells in the absence of any cell division. The proposed sequence of positive selection of class I MHC-restricted CD4-8+ T cells in the thymus[30] can be combined with our concept as follows.[18,19] Immature CD4+8+ thymocytes normally express a large variety of TCRs and are BGA-related glycoepitope positive. Some of the CD4 + 8 + cells TCRs of which can be bound to class I MHC ligands on thymic epithelium are rescued from cell death and selected to mature. Thus, nondividing CD4+8+ cells with high TCR levels and masked BGA-related glycodeterminants (possibily, by sialylation) are formed. Then they become nondividing CD4-8+ TCR high mature thymocytes with no expression of BGA-related glycoepitopes on their cell surface. At the stage of negative selection, T cells with high affinity receptors to self peptides-self MHC molecules become BGA-related glycoepitopes positive, for example, as a result of transfer of self peptides from the cell surface of antigen presenting cells into T cells and subsequent association of self peptides with MHC molecules of corresponding T cells. Thus, the T cells with high, medium and low levels of affinity of self peptide-self MHC complexes will have corresponding levels of association of self peptides to their own MHC and correspondingly, high, medium and low levels of BGA-related glycoepitope expression on their cell surface. Subsequently, T cells with high densities of BGA-related glycoepitopes on cell surfaces, as immature thymocytes, will be targets for elimination. This "self peptide pick-up mechanism" by T cells provides one of the possible explanations of how antigen-presenting function could participate in negative selection during thymic education and, probably, in positive selection. For example, association of T cell receptors (TCRs) with ligands could protect the former from proteolytic degradation and lead to TCR up-regulation. Subsequent up-regulation of MHC could be a second checking and selective signal for T cells with high affinities to MHC-self peptide complexes.

Desialylation of cell membrane glycomacromolecules could be a major factor in determining the activation of the cell adhesion mechanism. Despite their larger surface areas, T cell blasts and thymocytes have five-fold less sialic acid per cell than resting T cells[31] and are less negatively charged,[32] which may be important factors determining whether the adhesive mechanisms are active or latent. In lymph nodes, the activated antigen-responsive lymphocytes that aggregate in germinal centers are greatly undersialylated, whereas areas containing B and T cells in rapid transit between blood and lymph are sialylated normally.[33] Polysialylation of NCAM in the nervous system antagonizes its ability to promote adhesion.[34] Desialylation of the cell membrane glycomacromolecules could be a major mechanism of the regulation of negative charge on the cell surface, which is mainly responsible for preventing close cell-cell contact between circulating cells. Thus, a decrease in the sialylation level of the cell membrane glycomacromolecules could be a simple primary mechanism for the activation of cell adhesion. It would be accompanied by appropriate changes of physical properties (neutralization of negative charge) and structural characteristics (expression or exposition of cell adhesion glycodeterminants, particularly BGA-related glycoepitopes) of cell membrane, alteration of the morphology of contact between cells and activation of latent cell adhesion mechanisms. However, the carbohydrate-mediated cell-cell recognition regulatory mechanism may play an important role not only in initiation of multistep "cascade" cell adhesion reactions. It has been suggested

that the extent of glycosylation on ICAM-1 may regulate the endothelium adhesion to LFA-1 or MAC-1 (the leukocyte integrins),[35] and thus, control the stable binding of leukocyte to the endothelium, since this reaction involves the activation-dependent leukocyte integrin, MAC-1.[5]

Specific carbohydrate ligands are involved in the differential attachment to endothelium and subsequent migration into the tissues of specific lymphocyte subsets. Naive and memory-type T cells show clear differences in their distribution and migration patterns. The memory T cells accumulate in inflammatory lesions and are the only T cells which recirculate through uninflamed tissues including skin.[36] In contrast, the local lymph nodes are designed for the massive migration of lymphocytes of diverse specificities, especially naive T cells. Picker et al[37] and Shimizu et al[38] showed that a molecule termed endothelial cell-adhesion molecule-1 (ELAM-1) is one of the receptors on endothelium for a specific subset of circulating T lymphocytes: only memory-type T cells (CD4+ T cells), a functionally potent class of T cells, bind ELAM-1. ELAM-1 was originally implicated in the binding of blood neutrophils and monocyte-like cells to endothelium.[39] Ligands for ELAM-1 expressed on neutrophils are sialylated, fucosylated polylactosamines, such as the sialosyl Lewis X determinants, which bind to the lectin domain of selectins (see above). The carbohydrate ligand for ELAM-1 for T cells has not been identified, but it appears to be highly restricted in its expression to a specific subset of memory T cells, particularly of resting peripheral CD4 + T cells.[38] Since ELAM-1 has been identified as a molecule involved in the activation-independent binding of resting circulating memory T cells to endothelium, this adhesion pathway will allow inflamed endothelium to selectively "capture" that particular subset of lymphocytes.

L-selectin is constitutively functional, presents at high levels on circulating, non-activated, resting neutrophils and mediates their attachment to cytokine-stimulated endothelium by presenting neutrophil carbohydrate ligands, e.g., sialyl Lewis X, to the vascular E- and P-selectins.[40] However, in the absence of secondary activation-triggered stable attachment or in the presence of adhesion inhibitor(s) this primary adhesion is reversible and arrested circulating cells will be released and returned to the circulation.[5] Even activation-dependent adhesion via leukocyte integrins, that is stable for minutes under physiologic shear forces, is in principle reversible: neutrophils bound to endothelium by integrin-mediated adhesion (MAC-1) are released spontaneously after 10-15 minutes.[41] Thus, extravasation is not obligatory following even specific stable attachment of leukocytes to endothelium.

Therefore, specific interactions of selectins with corresponding carbohydrate ligands, particularly BGA-related glycodeterminants, are essential for the selective attachment of neutrophils, monocytes and lymphocytes to endothelium and their subsequent migration into the tissues—the process which is critically important for the maintenance of homeostasis. Furthermore, interaction of T cells with antigen-presenting cells also involves specific carbohydrate determinants, which are presumably associated with MHC molecules.

REFERENCES

1. Springer, T.A. (1990). Adhesion receptors of the immune system. Nature, 346:425-434.
2. Butcher, E.C. (1986) The regulation of lymphocyte traffic. Curr. Topics Microbiol. Immun. 128: 85-122.
3. Yednock, T.A., Rosen, S.D. (1989) Lymphocyte homing. Adv. Immun. 44:313-378.
4. Stoolman, L.M. (1989) Adhesion molecules controlling lymphocyte migration. Cell 56:907-910.
5. Butcher, E.C. (1991) Leukocyte-endothelial cell recognition: three (or more) steps to specificity and diversity. Cell 67:1033-1036.
6. Larsen, E., Palabrica, T., Sajer, S., Gilbert, G.E., Wagner, D.D., Furie, B.C., Furie, B. (1990). PADGEM-dependent adhesion of platelets to monocytes and neutrophils is mediated by a lineage-specific carbohydrate, LNF III (CD15). Cell, 63:467-474.
7. Phillips, M.L., Nudelman, E., Gaeta, F.C.A., Perez, M., Singhal, A.K., Hakomori, S., Paulson, J.C. (1990). ELAM-1 mediates cell adhesion by recognition of a carbohydrate ligand, sialyl-Lex. Science, 250:1130-1132.
8. Polley, M.J., Phillips, M.L., Wagner, E.A., Nudelman, E., Singhal, A.K., Hakomori, S., Paulson, J.C. (1991). CD 62 and endothelial cell-leukocyte adhesion molecule 1 (ELAM-1) recognize the same carbohydrate ligand, sialyl-Lewis X. Proc. Natl. Acad. Sci. USA, 88:6224-6228.
9. Berg, E.L., Robinson, M.K., Mansson, O., Butcher, E.C., Magnani, J.L. (1991). A carbohydrate domain common to both sialyl LeA and SialylLex is recognized by the endothelial cell leukocyte adhesion molecule ELAM-1. J. Biol. Chem., 266:14869-14872.
10. Lawrence, M.B., Springer, T.A. (1991). Leukocytes roll on a selectin at physiologic flow rates: distinction from and prerequisite for adhesion through integrins. Cell, 65:859-873.
11. Springer, T.A., Lasky, L.A. (1991) Sticky sugars for selectins. Nature 349:196-197.
12. Fukuda, M., Spooncer, E. S., Oates, J.E., Dell, A., Klock, J.C. (1984). Structure of sialylated fucosyl lactosaminoglycan isolated from human granulocytes. J. Biol. Chem., 259:10925-10935
13. Symington, F.W., Hedges, D.L., Hakomori, S.-I. (1985). Glycolipid antigens of human polymorphonuclear neutrophils and the inducible HL-60 myeloid leukemia line. J. Immunol., 134:2498-2506.
14. Glinsky, G.V. (1990). Immunoselective hypothesis of tumor progression. Role aberrant glycosylation, anticarbohydrate antibodies, extracellular glycomacromolecules and glycoamines. Journal of Tumor Marker Oncology. V.5, N 3, p. 206.
15. Glinsky, G.V. (1990). Glycoamines, aberrant glycosylation and cancer: a new approach to the understanding of molecular mechanism of malignancy., In: Molecular Oncology. Oncodevelopment proteins and clinical applications. XVIIIth meeting of the International Society for Oncodevelopmental Biology and Medicine. Abstract Book, Moscow, USSR, September 23-27, 1990, p.7.
16. Glinsky, G.V., Semyonova-Kobzar, R.A., Berezhnaya, N.M. (1990). Modification of cellular adhesion, metastasizing and immune response by glycoamines: Implication in the pathogenetical role and potential therapeutic application in tumoral disease. Journal of Tumor Marker Oncology., V.5, N 3, p. 231.
17. Glinsky, G.V. (1992). Glycoamines: Structural-Functional characterization of a new class of human tumor markers. In: Serological Cancer Markers. Editor: S. Sell. The Humana Press., Totowa, NJ, Chapter 11, p. 233-260.
18. Glinsky, G.V. (1992). The blood group antigens (BGA)-related glycoepitopes. A key structural determinants in immunogenesis and cancer pathogenesis. Critical Reviews in Oncology/Hematology, 12:151-166.
19. Glinsky, G.V. (1992). The blood group-related glycoepitopes: Key structural determinants in immunogenesis and AIDS pathogenesis. Medical Hypotheses (in press).
20. Hart, G.W. (1982). The role of asparagine-linked oligosaccharide in cellular recognition by thymic lymphocytes. J. Biol. Chem. 257:151-158.
21. Cullen, S.E., Kindle, C.S., Shreffler, D.C., Cowing, C. (1981). Differential glycosylation of murine B cell

and spleen adherent cell Ia antigens. J. Immunol. 127:1478-1484.

22. Cowing, C., Chapdelaine, J.M. (1983). T cells discriminate between Ia antigens expressed on allogeneic accessory cells and B cells: A potential function for carbohydrate side chains on Ia molecules. Proc. Natl. Acad. Sci. USA 80:6000-6004.

23. Taiara, S., Kakiuchi, T., Minotti, M., Nariuchi, H. (1986). The regulatory role of sialic acids in the response of class II reactive T cell hybridomas to allogenic B cells. J. Immunol. 137:2448-2454.

24. Powell, L.D., Whiteheart, S.W., Hart, G.W. (1987). Cell surface sialic acid influences tumor cell recognition in the mixed lymphocyte reaction. J. Immunol. 139:262-270.

25. Black, P., Vitetta, E., Forman, J., Kang, C.-Y., May, R., Uhr, J. (1981). Role of glycosylation in the H-2 restricted cytolysis of virus-infected cells. Eur. J. Immunol. 11:48-55.

26. Booy, C.J.P., Needjes, L.J., Boes, J., Ploegh, H.L., Melief, C.J.M. (1989) Specific immune responses restored by alteration in carbohydrate chains of surface molecules on antigen-presenting cells. Eur. J. Immunol. 19:537-542.

27. Adachi, M., Hayami, M., Kashiwagi, N., Mizuta, T., Ohta, Y., Gill, M.J., Matheson, D.S., Tamaoki, T., Shiozawa, C., Hakomori, S. (1988) Expression of LeY antigen in human immunodeficiency virus- infected human T cell lines and in peripheral lymphocytes of patients with acquired immune deficiency syndrome (AIDS) and AIDS-related complex (ARC). J. Exp. Med. 167:323-331.

28. Andrews, P.W., E. Gonczol, E., Fenderson, B.A., Holmes, E.H., O'Malley, G., Hakomori, S., Plotkin, S. (1989) Human cytomegalovirus induces stage-specific embryonic antigen 1 in differentiating human teratocarcinoma cells and fibroblasts. J. Exp. Med. 169:1347-1359.

29. Lloyd, K. O., Old, L.J. (1989). Human monoclonal antibodies to glycolipids and other carbohydrate antigens: dissection of the humoral immune response in cancer patients. Cancer Res. 49:3445-3451.

30. Huesmann, M., Scott, B., Kisielow, P., von Boehmer, H. (1991). Kinetics and efficacy of positive selection in the thymus of normal and T cell receptor transgenic mice. Cell 66:533-540.

31. Despont, J.P., Abel, C.A., Grey, H.M. (1975). Sialic acids and sialyltransferases in murine lymphoid cells: indicators of T cell maturation. Cell Immun. 17:487-494.

32. Shortman, K., von Boehmer, H., Lipp, J., Hopper, K. (1975). Subpopulations of T-lymphocytes. Transplant. Rev. 25:163-210.

33. Butcher, E.C., Rouse, R.V., Coffman, R.L., Nottenburg, C.N., Hardy, R.R., Weissman, I.L. (1982). Surface phenotype of Peyer's patch germinal center cells: implications for the role of germinal centers in B cell differentiation. J. Immun. 129:2698-2707.

34. Rutishauser, U., Acheson, A., Hall, A.K., Mann, D.M., Sunshine, J. (1988). The Neural Cell Adhesion Molecule (NCAM) as a Regulator of Cell-Cell Interactions. Science 240:53-57.

35. Diamond, M.S., Staunton, D.E., Marlin, S.D., Springer, T.A. (1991) Binding of the integrin Mac-1 (CDIIb/CD18) to the third immunoglobulin-like domain of ICAM-1 (CD54) and its regulation by glycosylation. Cell 65:961-971.

36. Mackay, C.R., Martson, W.L., Dudler, L. (1990). Naive and memory T cells show distinct pathways of lymphocyte recirculation. J. Exp. Med. 171:801-817

37. Picker, L.J., Kishimoto, T.K., Smith, C.W., Warnock, R.A., Butcher, E.C. (1991). ELAM-1 is an adhesion molecule for skin-homing T cells. Nature, 349:796-799.

38. Shimizu, Y., Shaw, S., Graber, N., Copal, T.V., Horgan, K.J., VanSeventer, G.A., Newman, W. (1991). Activation-independent binding of human memory T cells to adhesion molecule ELAM-1. Nature 349:799-802.

39. Bevilacqua, M.P., Pober, J.S., Mendrick, D.L., Cotran, R.S., Gimbrone, M.A. (1987). Identification of an inducible endothelial-leukocyte adhesion molecule. Proc. Natl. Acad. Sci. USA 84:9238-9242.

40. Picker, L.J., Warnock, R.A., Burns, A.R., Doerschuk, C.M., Berg, E.L., Butcher, E.C. (1991). The neutrophil selectin LECAM-1 presents carbohydrate ligands to the vascular selectins ELAM-1 and GMP-140. Cell 66:921-933.

41. Lo, S.K., Detmers, P.A., Levin, S.M., Wright, S.D. (1989). Transient adhesion of neutrophils to endothelium. J. Exp. Med. 169:1779-1793.

CHAPTER 5

THE ROLE OF BGA-RELATED GLYCODETERMINANTS IN CANCER:
PHENOTYPIC DIVERGENCE, CLONAL EVOLUTION, TUMOR PROGRESSION AND IMMUNOSELECTION

The capacity of cancer cells to invade adjacent tissue and to colonize distant sites is the fundamental feature of malignancy. The other important characteristics of cancer cells are the ability to escape from host growth control mechanisms with a tendency to increase tumor growth rate and "growth fraction"; loss of morphological and metabolic attributes of differentiation; extensive genetic, including chromosomal, aberrations; excessive accumulation and elaboration of several biologically active molecules such as hormones, growth factors, primary amines, glycomacromolecules, etc.; decreased antigenicity; and development of drug resistance. There is a considerable variability in time period and sequence in which these different features of cancer cells become apparent during tumor progression. It has been pointed out that this process appears to develop in a stepwise fashion through qualitatively different stages.[1,2] The concept of clonal evolution and clonal selection is generally accepted as a description of the mechanism of tumor progression on the cellular level.[3-8] This means that neoplasm arises from a single transformed cell, with the progeny of that cell expanding as a neoplastic "clone." Subsequent clonal evolution provides a basis for the heterogeneity of tumor cell population and sequential selection of variant subpopulations within this clone will result in the clinical, morphological, biochemical, and biological events of tumor progression. It has been suggested that enhanced genetic instability within the tumor cell population increases the probability of further genetic alterations, their subsequent selection, which provides a basis for clonal evolution of tumor. There is a growing evidence that neoplastic cells are more genetically unstable than comparable normal cells and that this genetic instability may be a major factor contributing to the phenomenon of clonal evolution.[8,9] However, there is no idea or concept how the genetic changes, which are very dissimilar in different types of cancer or even in different cases of the same type of cancer, will result in very close phenotypic aberrations that consequently cause the malignant social behavior of cancer cells.

Antigen Presentation and Tumor Progression

In some instances, this problem is very close to the problem of the relationship between the mutations and cancer. It has always unclear how such a broad spectrum of mutagens like viruses, different classes of chemicals, and radiation, that obviously cause different mutation patterns, may finally lead to the essentially same or very similar phenotypic alterations named malignant transformation and to the formation of full malignant phenotype through tumor progression. According to the concept that will be presented here, there is no specific genetic alterations or mutations required for formation of the full set of phenotypic abnormalities of cells typically associated with cancer. This concept identifies the antigen presenting function of an "abnormal" or "foreign" proteins as a major biological driving force that transfers genetic changes or mutations of any of a broad range of the intracellular proteins into the essentially same or very similar phenotypic changes of the cellular surface: expression and/or exposition of certain types of BGA-related glycodeterminants. This idea provides a basis for understanding the processes of phenotypic divergence, clonal evolution, and clonal selection of certain types of cancer cells through immunological mechanisms regardless of the nature and type of tumor. A main structural target of immunoselection by preexisting host immunological mechanisms, particularly, by naturally occurring anticarbohydrate antibodies, is the cancer cell with different densities of BGA-related glycoepitope on the cellular membrane. This concept outlines also the preservation of certain key features of malignancy, like autonomous proliferation and aberrant expression of BGA-related glycoepitopes, regardless of the type of cancer and stage of tumor progression. The other characteristics of malignancy may not be necessarily associated with the neoplastic state and the degree of their expression may vary quite independently, depending on the type of cancer, stage of progression or even the individual case of cancer.

It has been suggested that the host immune system represents one type of selective pressure through different cellular and humoral mechanisms on evolving neoplasms, particularly in the early stages.[10] Our concept postulates the key effector (naturally occurring anticarbohydrate antibodies) and target (BGA-related glycodeterminants) parts of this process. The hypothesis has been pointed out with regard to the monomolecular humorally mediated mechanism of the development of malignant behavior of cancer cells.[11-14] It has been suggested that the chronically proliferating tissue the high "growth fraction," the releases into extracellular medium an excessive amount of low molecular weight biologically active substances such as primary amines (putrescine, spermine, spermidine, histamine, etc.) and glycoamines. These chemically highly reactive compounds interact through nonenzymatic as well as enzymatic mechanisms with several long living proteins such as protease inhibitors, hormone- and growth factor-binding proteins and extracellular matrix proteins (collagen, fibronectin, laminin). Covalent modification of these long living proteins and subsequent alterations in their biological functions accumulate over time and form an aberrant extracellular molecular environment supporting malignant behaviour of cancer cells: invasion, dissemination from primary tumor, and colonization of distant organs.[11-14] This mechanism is relatively nonspecific and develops regardless of origin, type and individual characteristics of cancer as a consequence of the high level of cell multiplication. However, release by cancer cells into extracellular medium of carbohydrate-bearing macromolecules presumably would have a cancer cell type-specific character and would provide a basis for formation in extracellular medium and blood serum of a "humoral molecular imprint" of certain cancer membrane-associated glycoantigens. Since at least some of those glycoepitopes are involved in cell-cell recognition, association, and adhesion this event will contribute significantly to the modification of cancer cell adhesion.

The development of the human cancer is a long-term process, often spanning one-third to two-thirds of the lifetime of the individuals.[15] The prolonged sequence of events begining from the initial lesions (focal proliferations) and ending with the metastatic spread of cancer cells is encompassed by the term "tumor progression". In many cases of both experimental and clinical cancers, the initial lesions are numerous, progression to the next class of lesion is rare,[16] and the initial benign lesions are either reversible or may persist for extended period of time without progression. Cells are maintained in an orderly state in the organism by a hierarchy of organizational controls.[17] Progression to the malignant state in the organism may have the same basic principle of the state of selection as in the cell culture.[18] Soluble extracellular (serum) compounds such as primary amines (polyamines, glycoamines, etc.), serum glycomacromolecules, including numerous carbohydrate-associated cancer markers, glycoamines, etc., carbohydrate- recognition proteins, e.g, lectins, anticarbohydrate antibodies, etc., may well be an important chemical selection factors that significantly contribute to progressive malignant state selection and lead to the formation of full malignant phenotype. It is particularly important that this process may involve carbohydrate-mediated immunoselective mechanisms which will be described below. Serum factors are definitely involved in such a process because either progression or regression may occur in culture depending on the type of serum used in maintaining the cell during their passage.[18]

We have suggested[13,14,19-22] that the process of antigen presentation is accompanied by certain changes in the glycosylation pattern of the cellular membranes. These changes include the presentation of the BGA-related glycoepitopes—in particular, of their cryptic forms. Aberrant glycosylation may occur as a result of binding of foreign peptides to MHC and subsequent changes in the glycosylation pattern of complexesof peptide-MHC and/or as a result of presentation on the cell membrane of peptide-MHC complexes and exposition of new previously masked glycodeterminants. Experimental data are now available that confirm this suggestion for antigen presentation. (See Chapter 4.) Effective antigen presentation occurs in cells expressing an increased number of histocompatibility antigens with decreased sialylation.[96] The cells infected with AIDS virus or cytomegalovirus start to express a high level of LeY and LeX glycoepitopes, respectively.[97,98] This relatively nonspecific mechanism of the presentation of BGA-related glycoepitopes may be involved in the process of homotypic sorting of the cells at

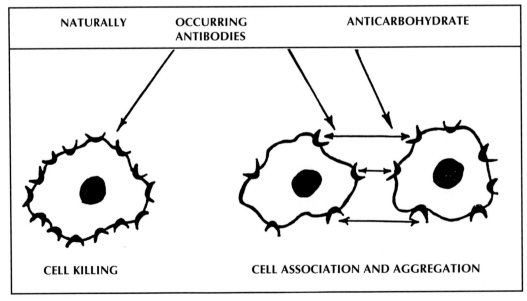

NATURALLY	OCCURRING ANTIBODIES	ANTICARBOHYDRATE

CELL KILLING   **CELL ASSOCIATION AND AGGREGATION**

Fig. 1. Hypothetical role of naturally occurring anticarbohydrate antibodies: Elimination of cells with high-density of BGA-related glycoepitopes and immunoselection of cells with medium- or low-density of BGA-related glycoepitopes.

subsequent stages of formation of the immune response. In view of this, natural anticarbohydrate antibodies may fulfill a double function: to accomplish the complement-dependent lysis of cells containing high-density BGA-related glycoepitopes, e.g., in thymic education it may lead to negative selection of T cells whose major histocompatibility complexes have a high affinity for their own presented antigens, and to favor the homotypic selection of cells revealing a medium or low density level of the BGA-related glycoepitopes.

Similar mechanisms may be realized for tumor cells, but these interactions should also involve humoral glycomacromolecules that contain BGA-related glycoepitopes. At present, most of the clinically used humoral tumor markers are glycomacromolecules (CEA, mucins, Ca 19-9, Ca15-3, CA 26); and many of them contain the BGA-related glycoepitopes (Lewis family, T, Tn, I)[76-81] and reveal immunosuppressive properties.[82] At least some of those proteins are directly involved in control of cancer cell adhesion probably through a carbohydrate-mediated mechanism. Recent studies suggest that CEA is a cellular adhesion molecule and may be potentially involved in cancer cell invasion and metastasis.[23-25] Biochemical evidence indicates that oligosaccharide side chains of the CEA molecules contain the LeX trisaccharide.[184] LeX or LeY antigens was found in carcinoembryonic antigen (CEA) purified from several different cases of colonic cancer. CEA of some cases carried the LeY epitope, some cases carried dimeric LeX.[99] The chemical structure of LeX/LeY-bearing carbohydrate chains at the peripheral region of N-linked glycan in CEA has been investigated in details.[81] Human chorionic gonadotropin, biological marker of early pregnancy and some types of cancer, which is involved in implantation and shows immunosuppressive activity in vitro and in vivo, also contains similar epitopes to CEA the BG-A like determinants of which do not have N-acetylgalactosamine.[185]

Blood serum from most healthy individuals contain natural anticarbohydrate antibodies, including those against the BGA-related glycoepitopes (anti-A, B, H, O, I, and Lewis family; anti-T and anti-Tn-antibodies; antiasialo GM_2 [GalNac β (1-4) Gal β (1-4) Glc-Cer]).[83] Blood serum from oncological patients reveals a decrease in the titers of anti-T, anti-Tn, anti-I and anti-Forssman anticarbohydrate antibodies.[58,76,84] Finally, the most characteristic manifestation of aberrant glycosylation of cancer cells is neosynthesis (or ectopic synthesis), the synthesis of incompatible antigens and incomplete synthesis (with or without the accumulation of precursors) of the BGA-related glycoepitopes.[85-87] BGA-related glycoepitopes are directly involved in the homotypic (tumor cells, embryonal cells) and heterotypic (tumor cells-normal cells) formation of cellular aggregates, e.g., Lewis X antigens; H-antigens, polylactosamine sequences; and T- and Tn-antigens, which was demonstrated in different experimental systems.[55-59] BGA-related alterations in the tissue glycosylation pattern are detected in benign (premalignant) tumors with high risk of malignant transformation, in primary malignant tumors, and in metastases,[87-93] i.e., they were demonstrated as typical alterations in different stages of tumor progression. It seems very intriguing to suggest that a similar mechanism is involved in the development of tumor progression and determines both its direction and the selection of metastasizing cell in cancer. Indirect confirmation of this idea is provided by two groups of facts: the preservation at all stages of tumor progression (from premalignant benign tumors with high risk of malignant transformation to metastatic foci) of the characteristic of primary malignant tumor alterations in the glycosylation pattern of BGA-related glycoepitopes of cellular membranes[87-93] and the accumulation during tumor progression of tumor cells with a low density level of histocompatibility antigens on the cellular membrane.[100]

Many human tumors, particularly those of epithelial origin and melanoma, appear to express greatly reduced levels of surface class I MHC molecules,[26] and the correlation has been found between MHC reduction, metastasis development, and poor prognosis. Those tumors include basal cell carcinoma,[50] squamouscell carcinoma,[52] eccrine porocarcinoma,[51] neuroblastoma[200] and small-cell lung carcinoma.[53] The small-cell lung carcinoma typically exhibits more rapid growth, poor prognosis, and earlier metastasis than other lung malignancies, where decreased levels of class I MHC expression was not observed. The relative expression of both class I and II HLA molecules on the cells correlates with malignancy of the melanoma. It has been reported that a) visceral metastases have a significantly lower class I expression than primary melanoma and locoregional metastasis;[54] b) patients with low class I expression on their melanoma cells had a significantly shorter survival than patients with a higher class I MHC expression on their tumors;[60] and c) the expression of HLA-Dr antigen in primary melanomas, espetially in the absence of DP, DQ, or ABC expression, is an adverse prognostic sign and is associated with high risk of metastasis.[61]

Finally, it has been suggested that tumor progression and invasiveness of melanoma correlate well with a decrease in class I MHC antigen expression and the appearance of class II antigen on melanoma cells.[54,62] In view of the role of carbohydrate in cell-cell recognition and adhesion, it is interesting to note that the structure of class II antigens on melanoma cells appears to be similar to that of antigens on lymphoid cells except for differences in glycosylation.[63] A decreased level of MHC expression compared to normal tissue was observed also in infiltrating ductal carcinomas[72,73] and mucinous colorectal carcinoma.[74] A significant correlation between loss of class I antigen expression and degree of morphological dedifferentiation was found in study of colorectal carcinoma.[75] A representative sample of the literature in this area has been summarized.[181]

It has been suggested that tumor antigens, including MHC antigens, may represent heritable alterations in cell-cell recognition sites, the dysfunction of which may result in abnormal or invasive growth.[27,28] MHC molecules have been described involving MHC-associated adhesion, homing, and contact inhibition.[29,30] Two major non-immunological functions of MHC molecules have been proposed: first, MHC antigens are involved as cell adhesion molecules in many cell-cell interactions;[64,65,67] second, the class I MHC molecules might be functionally associated with a variety of peptide hormone receptors, including insulin[70] glucagon,[65] epidermal growth factor,[68] γ-endorphin,[69] and therefore, MHC molecules may contribute to activation of transmembrane signaling pathway after cell-cell type specific adhesion. Barber et al[71] and Edidin[66] have proposed that free class I MHC chains can interact with hormone receptors as well as other self or nonself molecules in a manner similar to the association with β2M, thereby influencing many cellular functions. In some cases, however, the quantitative differences in HLA expression on cancer cells could be observed due to masking of the membrane-bound HLA-antigens by sialic acid glycoconjugates, since HLA expression on tumor cells could be detected after treatment of the cells with neuraminidase[182] and a highly sialylated mucin-like glycoprotein, epiglycanin, masks histocompatibility antigens on the mammary carcinoma line TA3-Ha to the extent that it can grow xenogeneically.[183]

Aberrant Glycosylation, Phenotypic Divergence of Tumor Cells, Immunosuppression, Immunoselection and Tumor Progression

Aberrant glycosylation of cell-membrane macromolecules is one of the universal phenotypic attributes of malignant tumors. A rather limited number of molecular probes based on monoclonal anticarbohydrate antibodies now enables the detection of over 90% of the most widespread human forms of cancer,[86,87] whereas the use of most informative kits of molecular probes for oncogene detection enables the detection of no more than 21-30% of human cancer.[101,102] It can be assumed that the conception of the oncogene that has significantly contributed to the understanding of molecular mechanisms of the control of cell proliferation and the automation of tumor-cell proliferation is, on the whole, of rather limited importance for the elucidation of mechanisms of the formation of other major biological properties of malignant cells. The concept of aberrant glycosylation undoudtedly is of more universal importance for the understanding of molecular mechanisms of the malignant behavior of transformed cells in vivo not only with egard to a classical approach to the knowledge of cellular and biological mechanisms of the formation of invasive and implantation properties, cell adhesion disturbances, and the like, but also for the elucidation of mechanisms of distant or systemic action of a malignant tumor on the homeostasis of host organism.

A number of theoretical and experimental preconditions implies that a malignant tumor is able to alter the glycosylation pattern of glycomacromolecules in the extracellular medium—in particular, in the blood serum. At least partly, these alterations may be characterized as the formation of humoral molecular imprints of glycomacromolecular antigens expressed on the membranes of tumor cells. One of the mechanisms of this process is evidently "shedding" of glycomacromolecules from the membrane of tumor cells.[103] Animal and human tumors have been extensively studied for their ability to shed cell-surface material. Alexander[188] and Kim et al[186] have proposed that the shedding of tumor-associated antigens allows tumor cells to escape host immune surveillance mechanisms and may play an important role in metastasis. Highly metastatic tumor cells shed cell-surface compounds, including tumor-associated antigens, at higher rates compared to cells of low metastatic potential.[186-189] Pellis and Kahan[190] were able to immunize animals successfully against cancer cell challenge with shed tumor-associated antigens. Recently, the successful immunotherapy of experimental cancer has been reported when immunization of animals was performed by synthetic[191] or purified natural[192] BGA-glycodeterminants. In view of the wide molecular range of biomolecules (from low molecular weight glycoamines to high molecular weight glycoproteins and multimolecular complexes) involved in these alterations and of the key role of the involved determinants in the development and maintenance of homeostasis, one can suggest that this mechanism is responsible for various distant (systemic) meta-

bolic and biochemical effects of malignant tumors on the organs and tissues of the host organism. This stage of the biochemical generalization of cancer evidently may be characterized as the early or first phase of the morphological generalization of the pathological process that is typical of cancer-metastasis development.

It has been suggested that one of the mechanisms of tolerance to the autoantigens may develop when autoantigens are shed in the circulation and are taken up by thymic macrophages or dendritic cells, which can then impart specific tolerance.[193] As we have already indicated, a number of tumor-associated carbohydrate antigens contain the specific BGA-related glycodeterminants that are also expressed on the surface of immune cells and are involved in leukocyte adhesion. Therefore, those tumor-associated carbohydrate antigens may interfere with or inhibit a number of physiological carbohydrate epitope-dependent leukocyte functions.

Specific MHC-unrestricted recognition of tumor-associated mucins by human cytotoxic T cells has been observed.[194] The mucin as a target antigen is atypical in its ability to directly bind and activate T cells in the absence of self MHC, presumably by abundant and regularly repeated peptide and carbohydrate antigenic epitopes. Therefore, T cells may recognize some of the tumor-associated carbohydrate antigens even without antigen-presenting cells. It has been suggested that production by cancer cells of high levels of circulating mucin, as is known to occur in patients with pancreatic and breast cancer,[195,196] may help tumor cells to escape immune destruction mechanism by engaging tumor-specific T-cell clones in the periphery with the soluble antigen.[194]

Bossman and Hall[197] found higher levels of β-galactosidase, α-mannosidase, and neuraminidase in malignant breast and colon tissues compared to surrounding normal tissues. Bossman et al[198] found higher levels of β-galactosidase, α-fucosidase, β-N-acetyl-galactosaminidase and β-N-acetylglucosaminidase activities in the high metastasis subline B16-F10 cells compared to the low metastasis subline B16-F1 cells. Four out of seven glycosidase activities (β-galactosidase, α-mannosidase, β-N-acetylgalactosaminidase and β-N-acetylglucosaminidase) were higher in a high metastatic subline of B16 melanoma cells compared to the low metastatic subline[199] but neuraminidase activity was lower. Therefore, a variety of oligosaccharide-degrading enzymes (glycosidase) show elevated levels in cancer tissues and this elevation has a correlation with metastatic potential of tumor cells. Tumor-associated carbohydrate antigens, particularly BGA-related glycoepitopes, may be released into extracellular medium by glycosidases and then interfere with homologous cellular glycodeterminants, according to the suggested earlier mechanism of nonenzymatic clustering of free glycodeterminants.[13]

When considering the causes of aberrant glycosylation of the glycomacromolecules of tumor cell membranes, one usually mentions the alterations in the glycosyltransferase activity, e.g., the activation of glycosyltransferase V, and less often, the elevation in the glyco-

sidase activity. In view of the important role of the primary structure of the protein carrier of carbohydrates in the determination of glycosylation specificity, it can be suggested that alterations in the expression of such glycocarrier molecules, e.g., ectopic in site and/or time expression, are also one of the causes of the changes in the glycosylation pattern of the membrane of tumor cells. In functional and biological aspects, the alterations in the glycosylation of glycomacromolecules of tumor-cell membranes can be related to the changes that reflect the automation and enhancement of cell division and also to the changes associated with blocking complete the development and differentiation program. The alterations associated with the automation of the division of transformed cells may be both the consequence of enhanced cell proliferation (since the entry of cells into mitotic cycle is accompanied by certain changes in the glycosylation of cellular membranes, and in tumor tissue the fraction of dividing cells is many orders of

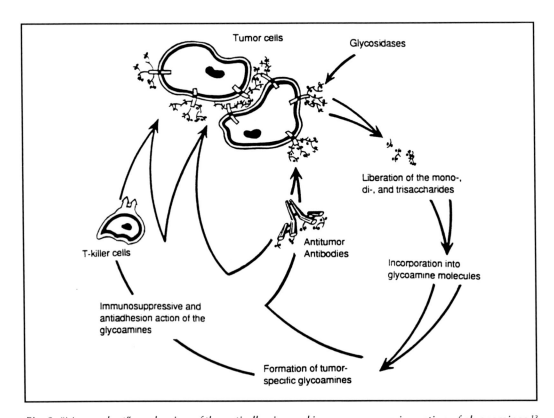

Fig. 2. "Monovalent" mechanism of the antiadhesive and immunosuppressive action of glycoamines.[13]

magnitude higher than that in the normal tissue) and the causal and/or supporting factors of the autonomous proliferation, e.g., the characteristic of transformation decrease in the level of GM_3-glycolipids of cellular membranes facilitates the dimerization of the EGF-receptor molecules.[104] However, in the biological aspects it seems to us most important that a part of the alterations in the glycosylation of glycomacromolecules of cellular membranes of tumor cells is associated with the antigen-presentation function of nonself peptides—a universal common biological function of all nucleus-containing cells of higher mammals and humans.

Mutations may create some variation in self peptides, but errors in gene expression, particularly translation, have been suggested as a possible source of major variabiligy.[105] The rate of misincorporation of amino acids, as estimated in bacterial and in vitro systems, ranges for 10^{-4} to 10^{-3} per amino acid.[106] Premature termination, a particular type of error, occurs in vivo in almost one-third of the 1000 amino acid long *Escherichia coli* β-galactosidase molecules.[107] In the absence of any special mechanism, a cell displaying some 10^6 MHC molecules on its surface would expose, at any time, a few hundred or a thousand distinct erroneous peptides. Their accumulation with time and with the large number of cells might develop a relatively large diversity: a total of order of 10^7 variants could be generated.[105]

The unavoidable noise of errors in gene expression, particularly translation, could be a major driving force in generation of erroneous peptides in tumor cells. In association with antigen-presentation function this mechanism may provide one the major causes of phenotypic divergence of tumor cells because association of erroneous peptides with MHC-molecules would be accompanied by abberrant glycosylation of MHC-glycoprotein-peptide complexes. In addition to the elevated (or even unchanged) genomic and mutation mutability of tumor cells, the antigen-presentation function and subsequent change in cellular surface with a definite type of the modification of the glycosylation pattern of cellular membrane glycoepitopes may provide the key cause of phenotypic mutability and divergence of tumor cells. These phenotypic alterations evidently may provide the immunoselective basis for tumor progression with a rather strict determination in tumoral disease of the selection direction, the latter largely depending on the predisposition of immunoselective factors of the host organism. Thus, the progression of tumor and the formation of complete malignant phenotype, including metastatic ability, may be represented as the consequence of a mono-pathogenetic immunoselective process with the following stages:

1. The automation of the proliferation of tumor cells;

2. Genomic mutability, and alterations in the primary structure of proteins;

3. Presentation of "altered peptides"—MHC complexes on cell surface, and aberrant glycosylation of membrane glycomacro-molecules; phenotypic divergence of tumor cells;

4. The formation of extracellular humoral "molecular imprints" of the glycoepitopes of glycomacromolecular antigens of cellular membranes of the tumor;

5. The immunoselection of tumorcell clones with a low or medium density of membrane glycoepitopes with competition for common structural determinants of the immunocompetent cells and humoral factors (anticarbohydrate antibodies, soluble glycomacromolecules and glycoamines); and

6. The formation of mechanisms of immunoresistance and homotypic association of tumor cells.

It seems very promising to consider as the key role the function in these processes of natural anticarbohydrate antibodies, glycoamines and other serum glycomacromolecules, and BGA-related glycoepitopes.

BGA-Related Glycoantigens in Embryogenesis, Organogenesis and Cell Differentiation

Glycoantigens undergo dramatic changes during development, cell differentiation and maturation.[85,108,109] We will only consider one example that clearly shows appearance, disappearance and reappearence of BGA-related glycoantigens on the cell surface in definite

stages of cell differentiation during embryogenesis, organogenesis, tissue formation and
maturation.

Cancer cells frequently express carbohydrate embryonic antigens since they undergo
a retrodifferentiation process during the course
of malignant transformation.[110] SSEA-1 (stage-
specific embryonic antigen-1) first described
by Solter and Knowles in 1978[111] is one of the
embryonic antigens which is specifically expressed on the murine preimplantation embryos at the morula stage. The SSEA-1 antigen
is described chemically as a carbohydrate antigen carrying LeX-hapten and i-antigenic
structures.[112,113] By using specific monoclonal
antibodies, it was shown that the SSEA-1 and
its modified form of the antigen, including
the sialylated form (sialylated LeX or sialylated
LeX-i) and fucosylated forms (LeY and poly
LeX), frequently accumulate in human cancer of various origins.[114-118] Lung adenocarcinoma is one of the good sources of the SSEA-1
and its modified forms of the antigen[119] and
murine monoclonal antibodies directed to
lung cancer cells frequently recognize SSEA-
1 and related antigens.[120,121] Moreover, among
the modified forms of SSEA-1, sialylated
SSEA-1 has been shown to be present in the
sera of patients with adenocarcinoma of the
lung, as well as in cancer cells.[117,122,123] Since
SSEA-1 was first described as an embryonic
antigen and subsequently found in various
human cancers, the antigen has sometimes
been regarded as an oncofetal or
oncodevelopmental antigen.[89,124-127]

The localization of three carbohydrate
antigens, LeX, LeY and sialylated LeX-i which
are closely related to stage-specific embryonic
antigen 1 in the lung of developing human
embryos was investigated using specific monoclonal antibodies.[110] The three antigens all
have physiological significance as stage-
specific developmental antigens of the human lung; those antigens were specifically
present in the bud cells at each important
step of the morphogenesis of the human lung,
such as cells in the lung buds, bronchial buds,
and terminal buds for the formation of the
alveolus, and cells differentiating into bronchial gland cells. The three antigens gradually disappear in the later stage of development

along with the maturation process of the lung.
Stage-specific embryonic antigen 1 and related antigens are known to be associated
with various human cancers, including lung
cancer. Miyake et al[110] suggest that the expression of these antigens in lung cancer cells
is the result of the retrodifferentiation of the
cancer cells to the stages of immature embryonic lung cells.

It was known that SSEA-1 antigens reappear in a certain lineage of cells after implantation during the course of morphogenesis of
certain organs such as brain and urogenital
organs,[126,128] and it was shown the stage specific expression of Le^x and related antigens in
developing lungs of human embryos.[110] One
type of expression of the stage-specific antigen is characterized by strict stage specificity
and disappearance of the antigen in mature
cells. This expression of SSEA-1-related antigen during the course of organogenesis has
been regarded as the second wave of the expression of those antigens[110] assuming the
appearance of these antigens in the preimplantation embryos as classically described[111]
as the first wave of the expression. The third
type of expression of stage-specific antigen is
characterized by the persistence of the antigen after differentiation. This expression of
SSEA-1 related antigens has been regarded as
the third and the latest wave of the expression
of these antigens in intrauterine life.[110]

Direct evidence for the involvement of
LewisX antigen in embryonic development
has been shown by studying the effects of
oligosaccharides on preimplantation-stage
mouse embryos. During compaction, blas-
tomeres maximize their intercellular contacts
and generate polar distribution of microvilli.
The appearance of SSEA-1 (LeX) on the mouse
embryo surface correlates in time with the
onset of compaction.[111] It has been suggested
and confirmed that the stage-specific appearance of LeX may play a role in stabilizing
compaction and in sorting LeX-positive blas-
tomeres to the inner cell mass during blasto-
cyst formation[31,32,56] and in the attachment of
mouse blastocysts on endometrial monolayer.[57]
The oligosaccharides purified from human
milk were tested in this experimental system
when added either as monomers or as multi-

valent haptens attached to the primary amino groups of lysyllysine.[31,56] Multivalent LeX haptens were found to make a dramatic decompaction of fully compacted 16-cell embryos, whereas multivalent LeA or GlcNac chitobiose haptens did not cause decompaction. It has been suggested that LeX valency is critically important, since a multivalent LeX hapten was able to decompact embryos, whereas free oligosaccharide was largely ineffective.[31,56] In many other biological systems carbohydrate recognition requires high density or clustering of sugar residues. For example, LeX is carried by the early mouse embryo on highly branched lactosaminoglycans (embryoglycans). Lactosaminoglycans are carriers for many stage-specific embryonic antigens, including SSEA-1 (LeX), LeY, I, i, and the ABH histo-blood group system.[56] However, lactosaminoglycans purified from F9 embryonal carcinoma cells caused rapid agglutination of mouse embryos,[56] agglutinate F9 embryonal carcinoma cells and participate in embryonal carcinoma cell adhesion.[33] Thus, the difference in the range of molecular weight of carrier as well as the density and valence of oligosaccharide haptens may cause the opposite biological effect mediated by the same glycodeterminant. It is important that as in the many other cases of cell-cell type specific recognition and adhesion the process of compaction requires the cooperation of multiple cell adhesion molecules and mechanisms. Compaction is known to be controlled by a number of cell surface molecules, including uvomorulin,[34] galactosyltransferase[35] and gap junction proteins.[36]

The Expression of BGA-Related Glycodeterminants on Cancer Cells.

As we already have mentioned, abberrant glycosylation of cell membrane macromolecules is one of the universal attributes of malignant tumors. A rather limited number of molecular probes based on monoclonal anticarbohydrate antibodies now enables the detection of over 90% of the most widespread forms of human cancer.[86,87] The most characteristic manifestation of aberrant glycosylation of cancer cells is neosynthesis (or ectopic synthesis), the synthesis of incom-

patible antigens, and imcomplete synthesis (with or without the accumulation of precursors) of the BGA-related glycoepitopes.[85-87]

T- and Tn-antigens

Springer and his coworkers have found that the blood group precursor antigens T and Tn are carcinoma-associated.[58] They have shown them to be unmasked on the external surface membranes of most primary carcinomas and their metastases. In all other tissues T- and Tn-antigens are masked and are generally precursors in normal complex carbohydrate chains. T and Tn have not been detected in benign tumors nor in other diseased or healthy tissues.[129-141] Although T and Tn antigens are present in most tissues, they are however, occluded by covering carbohydrate structures.[142,143] This is valid throughout life;[144-147] however, there is some evidence that T and Tn are stage-specific differentiation antigens expressed transiently on human fetuses.[148] All humans have pre-existing antibodies against T and Tn.[144-147] Human anti-T antibodies in the presence of complement kill carcinoma cells in vitro;[149] obviously, they do not kill carcinomas in vivo, at least not those that become clinically manifested. Anti-T IgM constitutes 7-14% of total IgM and its concentration was about 2.5 and about 5 times higher than that of anti-T IgG and anti-T IgA, respectively.[150,151]

T- and Tn-antigens are expressed in immunoreactive form in about 90% of carcinoma tissues.[58,95] T-antigen has been found in all colon and in 96% of breast carcinomas examined.[131-135] Springer et al found T-antigen in 95% of 144 fresh surgical samples of all types of primary carcinomas from all major organs and all seven metastatic lesions from four patients up to 6.5 years after removal of their T and Tn positive primary breast carcinoma strongly expressed these antigens.[129,141,152-154] Tn-antigen occurred in carcinomas about as frequently as T. Both of these antigens have been detected on the outer cell membranes of 25 of 26 human breast carcinoma-derived epithelial cell lines.[134,135,152] The authencity of the T- and Tn-active structures have been demonstrated immunochemically in antibody inhibition studies with specific haptens;[155-158] by a solid-phase immunoassay;[150,151] and biochemically with glycosidases[155] and with glycosyltransferases.[159-161] Thus, the unmasked T- and Tn-specific epitopes are unique carcinoma markers, particularly for breast and lung carcinomas. T-, Tn- and sialosyl-Tn structures have been identified as human tumor-associated carbohydrate antigens.[58,85,87,92,95,162-166] The expression of these antigens is highly specific to a variety of human cancers, and is essentially absent in normal tissues.[85,87,92,166] The high, specific inhibition activity of tumor cell adhesion to hepatocytes by T- and Tn-specific glycoconjugates has been observed.[58,59] Elevated serum level of T-antigen in patients with breast cancer has been observed[167] and monoclonal anti-T antibody has been successfully used for in vivo radioimmunoimaging of primary and metastatic carcinoma both in animal models and human cancer, particularly to localize in vivo metastatic cancer during of an extended phase I clinical trials.[167]

Thus, numerous independent lines of evidence have strongly supported the concept that T-antigen is expressed in cryptic form on rodent and human red blood cells and leukocyte (lymphocytes, monocytes, granulocytes) and is expressed in noncryptic or unsubstituted form on human carcinoma cells. The confirmatory data have been obtained on carbohydrate inhibition studies of the naturally occurring human antibody to neuraminidase-treated human lymphocytes;[37] the demonstration of the T-antigen on neuraminidase-treated human red blood cells and lymphocytes using synthetic hapten and polyclonal rabbit and human anti-T-antibodies;[158] the expression of T-antigen on a murine lymphomas, on a spontaneous murine mammary carcinoma, and on human adenocarcinoma has been observed using monoclonal antibody against a cryptic carbohydrate antigen of murine and human lymphocytes;[40] the evidence of T-antigen expression on human carcinoma cells as well as on lymphocytes, monocytes, and granulocytes has been shown using human lung carcinoma monoclonal antibody specific for the Thomsen-

Friedenreich antigen[95] and using monoclonal antibody against synthetic T-antigen.[38]

GALα (1-3) GAL EPITOPE

Anti-Gal antibody is a natural IgG antibody present in unusually large amounts in the serum of healthy individuals and constitutes 1% of circulating IgG.[168] Anti-Gal was found to be a polyclonal monospecific antibody interacting specifically with an oligosaccharide residue with the structure Galα (1-3) Galβ (1-4) GlcNac - R.[169,170] Anti-Gal is produced throughout life in humans as a result of what seems to be a constant antigenic stimulation by gastrointestinal bacteria.[171] A striking reciprocal evolutionary pattern in the distribution of anti-Gal and the Galα (1-3) Galβ (1-4) GlcNac- R residue was observed in mammals.[172,173] The Galα (1-3) Galβ (1-4) GlcNac - R epitope is abundant on various nucleated cells and red blood cells from nonprimate mammals, prosimians and New World monkeys, however, their expression is diminished on cells of Old World monkeys, apes, and humans. In contrast, anti-Gal antibody is present in the serum of Old World monkeys, apes and man, but this antibody is absent from the blood of New World monkeys and nonprimate animals.[172,173]

Expression of Galα (1-3) Gal cell surface epitpopes has been correlated with the metastatic potential of murine tumor cells. It has been demonstrated that there are more Galα (1-3) Gal residues in metastatic murine tumor cells than in normal or nonmetastatic or low-metastatic tumor cells.[174-177] Galα (1-3) Gal residues are expressed at the cell surface of malignant human cancer cells, including four cell lines and 50% of the malignant breast specimens obtained by aspiration biopsy.[178] In contrast, all benign breast biopsies and normal cells were Galα (1-3) Gal negative. Affinity-purified anti-Gal antibody significantly inhibited tumor cell attachment in two in vitro assays, attachment to perfused human umbilical vein endothelium, and attachment to isolated laminin.[178] Since anti-Gal antibody is a natural component of all human sera, it has been suggested that it may be part of the natural antitumor defense system in humans.[178] However, these antibodies obviously do not protect against clinically manifested breast carcinoma since at least 50% of malignant breast specimens are expressed $Gal\alpha$ (1-3) Gal epitope. The $Gal\alpha$ (1-3) Gal epitope occurs on normal human erythrocytes in small amounts and in a cryptic form, indicating that the suppression of synthesizing the $Gal\alpha$ (1-3) Gal epitope is not absolute.[168] All human sera contain a naturally occurring antibody which recognizes the $Gal\alpha$ (1-3) Gal epitope (see above). This antibody binds to human senescent, thalassemic and sickle erythrocytes having the exposed $Gal\alpha$ (1-3) Gal epitope, and leads to immune-mediated destruction of these cells.[173,179,180] The similar mechanism may lead to the immunoselection of metastatic clone of cancer cells during the tumor progression.

LEWIS FAMILY OF GLYCOANTIGENS

Stage specific embryonic antigen 1 and related antigens are known to be associated with various human cancers, including lung cancer. Cancer cells frequently express carbohydrate embryonic antigens since they undergo a retrodifferentiation process during the course of malignant transformation.[110] SSEA-1 (stage-specific embryonic antigen-1) first described by Solter and Knowles in 1978,[111] is one of the embryonic antigens which is specifically expressed on the murine preimplantation embryos at the morula stage. SSEA-1 antigen is described chemically as a carbohydrate antigen carrying Le^x-hapten and i-antigenic structures.[112,113] By using the specific monoclonal antibodies, it was shown that SSEA-1 and its modified form of the antigen, including the sialylated form (sialylated Le^x or sialylated Le^x-i) and fucosylated forms (Le^y and poly Le^x), are frequently accumulated in human cancer of various origins.[114-118] Lung adenocarcinoma is one of the best sources of SSEA-1 and its modified forms of the antigen,[119] and murine monoclonal antibodies directed to lung cancer cells frequently recognize SSEA-1 and related antigens.[120,121] Moreover, among the modified

forms of SSEA-1, sialylated SSEA-1 has been shown to be present in the sera of patients with adenocarcinoma of the lung as well as in cancer cells.[117,122,123] Since SSEA-1 was first described as an embryonic antigen and subsequently found in various human cancers, the antigen has sometimes been regarded as an oncofetal or oncodevelopmental antigen.[124-127]

Miyake et al[110] suggest that the expression of these antigens in the lung cancer cells is the result of the retrodifferentiation of the cancer cells to the stages of immature embryonic lung cells. The widespread expression of LeY antigen in epithelial cancers, particularly of nonsecretor individuals, has therapeutic potential[94] since these individuals do not synthesize LeB and LeY antigens in normal tissues or have only low levels in a few specific sites. Thus, the expression of these antigens is highly tumor-restricted in nonsecretor individuals. On the other hand, evidence has been presented by numerous investigators that the Lewis family of glycodeterminants are directly involved in cell-cell recognition, association and adhesion of histogenetically broad range of embryonal, lymphoid and cancer cells.[31,32,41-49,55-57]

Therefore, the aberrant glycosylation in cancer is characterized by expression on the cell surface of tumor cells of certain BGA-related glycodeterminants. These changes were demonstrated as typical for different stages of tumor progression, including metastasis. The BGA-related glycodeterminants that are expressed on the surface of cancer cells function as cell adhesion molecules. They are present in cancer blood serum in biologically active form and may either stimulate or inhibit cell-cell interactions. The important fact is that in serum of all normal individuals circulate the naturally occurring anticarbohydrate antibodies of the same specificity.

REFERENCES

1. Foulds, L. (1957) Tumor progression. Cancer Res., 17:355-356.
2. Foulds L. (1975) Neoplastic development, Vol. 2, New York: Academic Press.
3. Cairns, J. (1975) Mutation, selection and the natural history of cancer. Nature, 255: 197-200.
4. Nowell, P. (1976) The clonal evolution of tumor cell

populations. Science, 194:23-28.

5. Klein, G. (1979) Lymphoma development in mice and human: diversity of initiation is followed by convergent cytogenetic evolution. Proc. Natl. Acad. Sci. USA, 76:2442-2446.

6. Fialkow, P. (1979) Clonal origin of human tumors. Annu. Rev. Med., 30:135-143.

7. Arnold, A., Cossman, J., Bakhshi, A., Jaffe, E.S., Waldmann, T.A., Korsmeyer, S.J. (1983) Immunoglobulin gene rearrangements as unique clonal markers in human lymphoid neoplasms. N. Engl. J. Med., 309:1593-1599.

8. Nowell, P.C. (1986) Mechanisms of tumor progression. Cancer Res., 46:2203-2207.

9. German, J. (ed.) (1983) Chromosome Mutation and neoplasia. New York; Alan R. Liss.

10. Klein, G., Klein, E. (1985) Evolution of tumors and the impact of molecular oncology. Nature, 315:190-195.

11. Glinsky, G.V. (1987) J. Tumor Marker Oncology, 1:249-294.

12. Glinsky, G.V. (1989) Glycoamines: Biochemistry of a new class of humoral tumor markers. J. Tumor Marker Oncology, 3:193-221.

13. Glinsky, G.V. (1992) Glycoamines: Structural-functional characterization of a new class of human tumor markers. In Serological cancer markers. Editor: S. Sell. The Humana Press, Totowa, N.J., Chapter 11, pp. 233-260.

14. Glinsky, G.V. (1992) The blood group antigen (BGA)-related glycoepitopes. A key structural determinants in immunogenesis and cancer pathogenesis. Critical Reviews in Oncology/Hematology, 12:151-166.

15. Farber, E., Rubin, H. (1991) Cellular adaptation in the origin and development of cancer. Cancer Res. 51:2751-2761.

16. Clark, W.H. (1991) Tumour progression and the nature of cancer. Br. J. Cancer, 64:631-644.

17. Rubin, H. (1990) The significance of biological heterogeneity. Cancer Metastasis Rev., 9:1-20.

18. Rubin, H. (1992) Adaptive evolution of degrees and kinds of neoplastic transformation in cell culture. Proc. Natl. Acad. Sci. USA, 89:977-981.

19. Glinsky, G.V. (1990) J. Tumor Marker Oncology. 5:206.

20. Glinsky, G.V. (1990) In: Molecular Oncology. Oncodevelopment proteins and clinical applications. XVIIIth meeting of the International Society for Oncodevelopmental Biology and Medicine. Abstract Book, Moscow, USSR, September 23-27, 1990, p.7.

21. Glinsky, G.V., Semyonova-Kobzar, R.A., Berezhnaya, N.M. (1990) J. Tumor Marker Oncology., 5:231.

22. Glinsky, G.V. (1992) The blood group antigen-related glycoepitopes: Key structural determinants in immunogenesis and AIDS pathogenesis. Medical Hypotheses (in press).

23. Benchimol, S., Fuks, A., Jothy, S., Beauchemin, N., Shirota, K., Stanners, C.P. (1989) Carcino-embryonic antigen, a human tumor marker, functions as

24. Jessup, J.M., Giavazzi, R., Campbell, D., Cleary, K., Morikawa, K., Fidler, I.J. (1988) Growth potential of human colorectal carcinoma in nude mice: association with the preoperative serum concentration of carcinoembryonic antigen in patients. Cancer Res. 48:1689-1692.

25. Hostetler, R.B., Augustus, L.B., Mankarious, R., Chi, K.F., Fan, D., Toth, C. et al. (1990) Carcinoembryonic antigen as a selective enhancer of colorectal cancer metastasis. J. Natl. Cancer Inst. 82:380-385.

26. Gopas, J., Rager-Zisman, B., Bar-Eli, M., Hammerling, G.J., Segal, S. (1989) The relationship between MHC antigen expression and metastasis. Adv. Cancer Res., 53:89-115.

27. Boyse, E.A. (1970) In Immunosurveillance. (R.T. Smith and M. Landy, eds.), Academic Press, New York, pp. 5-48.

28. DeBaetseiler, P., Katzav, S., Gorelik, E., Feldman, M., Segal, S. (1980) Differential expression of H-2 gene products in tumour cells in associated with their metastatogenic properties. Nature, 288:179-181.

29. Dausset, J., Contu, L. (1980) Is the MHC a general self-recognition system playing a major unifying role in an organism? Hum. Immunol., 1:5-57.

30. Scofield, V.L., Schlumpberger, J.M., West, L.A., Weissman, I.L. (1982) Protochordate allorecognition is controlled by a MHC-like gene system. Nature, 295:499-501.

31. Fenderson, B.A., Zehavi, U., Hakomori, S. (1984) A multivalent lacto-N-fucopentaose III-lysyllysine conjugate decompacts preimplantation mouse embryos, while the free oligosacharide is ineffective. J. Exp. Med., 160:1591-1596.

32. Bird, J.M and S.J. Kimber (1984) Oligosaccharides containing fucose linked α (1-3) and α (1-4) to N-acetylglucosamine cause decompaction of mouse morulae. Dev. Biol., 104:449.

33. Shur, B.D. (1983) Embryonal carcinoma cell adhesion: The role of surface galactosyltransferase and its 90K lactosaminoglycan substrate. Dev. Biol., 99:360-372.

34. Takeichi, M. (1987) Cadherins: A molecular family essential for selective cell-cell adhesion and animal morphogenesis. Trends. Gen. 3:213-217.

35. Bayna, E.A., Shaper, J.H., Shur, B.D. (1988) Temporally specific involvement of cell surface $\beta 1$, 4 galactosyltransferase during mouse embryo morula compaction. Cell, 53:145-157.

36. Lee, S., Gilula, N.B., Warner, A.E. (1987) Gap junctional communication and compaction during preimplantation stages of mouse development. Cell, 51:851-860.

37. Rogentine, G.N., Plocinik, B.A. (1974) Carbohydrate inhibition studies of the naturally occurring human antibody to neuraminidase-treated human lymphocytes. J. Immunol., 113:848-888.

38. Longenecker, B.M., Williams, D.J., MacLean, G.D., Selvaraj, S., Suresh, M.R., Noujaim, A.A. (1987) Monoclonal antibodies and synthetic tumor-associ-

ated glycoconjugates in the study of the expression of Thomsen-Friedenreich-like and Tn-like antigens on human cancers. J. Natl. Cancer Inst., 78:489-496.

39. Bray, J., Lemiux, R.U., McPherson, T.A. (1981) Use of a synthetic hapten in the demonstration of the Thomsen-Friedenreich (T) antigen on neuraminidase-treated human red blood cells and lymphocytes. J. Immunol., 126:1966-1969.

40. Longenecker, B.M., Rachman, A.F.R., Leigh, J., Purser, A., Greenberg, A.H., Willans, D.J., Keller, O., Petrik, P.K., Thay, T.Y., Suresh, M.R., Noujaim, A.A. (1984) Monoclonal antibody against a cryptic carbohydrate antigen of murine and human lymphocytes. 1. Antigen expression in non-cryptic or unsubstituted form in certain murine lymphomas, on a spontaneous murine mammary carcinoma and on several human adenocarcinomas. Int. J. Cancer, 33:123-129.

41. Eggens, I., Fenderson, B., Toyokuni, T., Dean, B., Stroud M., Hakomori, S.I. (1989) Specific interaction between Le^x and Le^x determinants: A possible basis for cell recognition in preimplantation embryos and in embryonal carcinoma cells, J. Biol. Chem. 264:9476-9484.

42. Miyake, M., Hakomori, S.I. (1991) A specific cell surface glycoconjugate controlling cell motility: evidence by functional monoclonal antibodies that inhibit cell motility and tumor cell metastasis. Biochemistry, 30:3328-3334.

43. Springer, T.A., Lasky, L.A. (1991) Sticky sugars for selectins. Nature, 349:196-197.

44. Larsen, E., Palabrica, T., Sajer, S., Gilbert, G.E., Wagner, D.D., Furie, B.C., Furie, B. (1990) PADGEM-dependent adhesion of platelets to monocytes and neutrophils is mediated by a lineage-specific carbohydrate, LNF III (CD15) Cell, 63:467-474.

45. Phillips, M.L., Nudelman, E., Gaeta, F.C.A., Perez, M., Singhal, A.K., Hakomori, S., Paulson, J.C. (1990) ELAM-1 mediates cell adhesion by recognition of a carbohydrate ligand, sialyl-Lex. Science, 250:1130-1132.

46. Polley, M.J., Phillips, M.L., Wagner, E.A., Nudelman, E., Singhal, A.K., Hakomori, S., Paulson, J.C. (1991) CD 62 and endothelial cell-leukocyte adhesion molecule 1 (ELAM-1) recognize the same carbohydrate ligand, sialyl-Lewis X. Proc. Natl. Acad. Sci. USA, 88:6224-6228.

47. Berg, E.L., Robinson, M.K., Mansson, O., Butcher, E.C., Magnani, J.L. (1991) A carbohydrate domain common to both sialyl LeA and SialylLex is recognized by the endothelial cell leukocyte adhesion molecule ELAM-1. J. Biol. Chem., 266:14869-14872.

48. Hakomori, S.-I. (1991) Possible functions of tumor-associated carbohydrate antigens. Current Opinion in Immunology, 3:646-653.

49. Hakomori, S.-I. (1992) Possible new directions in cancer therapy based on aberrant expression of glycosphingolipids: Anti-adhesion and ortho-signalling therapy., Cancer Cells (in press).

50. Holden, C.A., Sanderson, A.R., MacDonald, D.M. (1983) J. Am. Acad. Dermatol., 9:867-871.

51. Holden, C.A., Shaw, M., McKee, P.H., Sanderson, A.R., MacDonald, D.M. (1984) Loss of β$_2$ microglobulin from the cell surface of cutaneous malignant and premalignant lesions. Arch. Dermatol., 120:732-735.

52. Turbitt, M.L., Mackie, R.M. (1981) Loss of membrane B2 microglobulin in eccrine porocarcinoma. Its association with the histopathologic and clinical criteria of malignancy. Br. J. Dermatol., 104:507-513.

53. Doyle, A., Martin, J., Fune, K., Gazdat, A., Carney, D., Martin, S., Linnoila, I., Cuitta, F., Mulshine, J., Bunn, P., Minna, J. (1985) Markedly decreased expression of class I histocompatibility antigens, protein, and mRNA in human small-cell lung cancer. J. Exp. Med., 161:1135-1151.

54. Broecker, E.B., Suter, L., Bruggen, J., Ruiter, D.J., Macher, E., Sorg, C. (1985) Int. J. Cancer, 36:29-35.

55. Fenderson, B.A., Andrews, P.W., Nudelman, E., Clausen, H., Hakomori, S.I. (1987) Glycolipid core structure switching from globo to lacto-and ganglio-series during retinoic acid-induced differentiation of TERA-2-derived human embryonal carcinoma cells., Dev. Biol. 122, 21-34.

56. Fenderson, B.A., Eddy, E.M., Hakomori, S.I. (1990) Glycoconjugate expression during embryogenesis and its biological significance. Bio Essays 12, 173-79.

57. Lindenberg, S., Sundberg, K., Kimber, S.J., Lundblad, A. (1988) The milk oligosaccharide, lacto-N-fucopentaose l, inhibits attachment of mouse blastocysts on endometrial monolayers., J. Reprod. Fert. 83, 149-158.

58. Springer, G.F. (1984) T and Tn, general carcinoma autoantigens. Science, 224, 1198-1206.

59. Springer, G.F., Cheinsong-Popov, R., Schirrmacher, V., Desoi, P.R., Tegtmeyer, H. (1983) Proposed molecular basis of murine tumor cell-hepatocyte interaction. J. Biol. Chem. 258, 5702-5706.

60. van Duinen, S.G., Ruiter, D.J., Broecker, E.B., van der Velde, E.A., Sorg, C., Welvaart, K., Ferrone, S. (1988) Level of HLA antigens in locoregional metastases and clinical course of the disease in patients with melanoma. Cancer Res., 48:1019-1025.

61. D'Alessandro, G., Zardawi, I., Grace, J., McCarthy, W.H., Hersey, P. (1987) Immunohistological evaluation of MHC class I and II antigen expression on nevi and melanoma: Relation to biology of melanoma. Pathology, 19:339-346.

62. Fossati, G., Anichimi, A., Taramelli, D., Balsari, A., Gambacorti-Passerini, C., Kirkwood, J.M., Permiani, G. (1986) Immune response to autologous human melanoma: implication of class I and II MHC products. Biochem. Biophys. Acta., 865:235-251.

63. Alexander, M.A., Hubbard, S.C., Strominger, J.L. (1984) HLA-DR antigens of autologous melanoma and B lymphoblastoid cell lines: differences in glycosylation but not protein structure. J. Immunol., 133:315-320.

64. Edidin, M. (1983) Immunol. Today, 4:269-270.

65. Edidin, M. (1986) Major histocompatibility complex haplotypes and the cell physiology of peptide hormones. Hum. Immunol., 15:357-365.

66. Edidin, M. (1988) Function by association? MHC antigens and membrane receptor complexes. Immunol. Today, 9:218-219.

67. Scofield, V.L., Schlumpberger, J.M., Weissman, I.L. (1982) Am. Zool., 22:783-794.

68. Schreiber, A.B., Schlessinger, J., Edidin, M. (1984) Interaction between major histocompatibility complex antigens and epidermal growth factor receptors on human cells. J. Cell Biol., 98:725-731.

69. Claas, F.H.J., van Ree, J.M., Verhoeven, W.M.A. (1986) The interaction between gamma-type endorphins and HLA class I antigens. Hum. Immunol., 15:347.

70. Simonsen, M., Skjodt, K., Crone, M. (1985) Compound receptors in the cell membrane: ruminations from the borderland of immunology and physiology. Prog. Allergy, 36:151-176.

71. Barber, B., Smith, M.H., Allen, H., Williams, D.B. (1988) In MHC + X: complex formation and antibody induction (P. Ivanyi, ed.) Springer-Verlag, Berlin and New York.

72. Natali, P.G., Giacomini, P., Bigotti, A., Imai, K., Nicotra, M.R., Ng, A.K., Ferrone, S. (1983) Heterogeneity in the expression of HLA and tumor-associated antigens by surgically removed and cultured breast carcinoma cells. Cancer Res., 43:660-668.

73. Natali, P., Bigotti, A., Cavalieri, R., Nicotra, M.R., Tecce, R., Manfredi, D., Chen, Y.-X., Nadler, L.M., Ferrone, S. (1986) Gene products of the HLA-D region in normal and malignant tissues of nonlymphoid origin. Hum. Immunol., 15:220-233.

74. Van den Ingh, H.F., Ruiter, D.J., Griffioen, G., van Muijen, G.N.P., Ferrone, S. (1987) HLA antigens in colorectal tumours—low expression of HLA class I antigens in mucinous colorectal carcinomas. Br. J. Cancer, 55:125-130.

75. Momburg, F., Degener, T., Bacchus, E., Moldenhauer, G., Hammerling, G.J., Moller, P. (1986) Loss of HLA-A,B,C and de novo expression of HLA-D in colorectal cancer. Int. J. Cancer, 37:179-184.

76. Dube, V.E. (1987) The structural relationship of blood group-related oligosaccharides in human carcinoma to biological function: a perspective. Cancer Metastasis Rev. 6, 541-565.

77. Kannagi, R., Hakomori, S.I., Imura, H. (1988) In: Altered glycosilation in tumor cells. Editors: Ch. L. Reading, S.I. Hakomori, D.M. Marcus., New York: Alan R. Liss, 279-94.

78. Linsley, P.S., Brown, J.P., Magnani, J.L., Horn, D. (1988) Monoclonal antibodies reactive with mucin glycoproteins found in sera from breast cancer patients. Cancer Res., 48, 2138-49.

79. Magnani, J., Nilsson, B., Brockhaus, M., Zop, J.D., Steplewski, Z., Koprowski, H., Ginsburg, V. (1982) A monoclonal antibody-defined antigen associated with gastrointestinal cancer is a ganglioside containing sialylated lacto-N-fucopentaose II. J. Biol. Chem.

257, 14365-14369.

80. Magnani, J., Steplewski, Z., Koprowski, H., Ginsburg, V. (1983) Identification of the gastrointestinal and pancreatic cancer-associated antigen detected by monoclonal antibody 19-9 in the sera of patients as a mucin. Cancer Res., 43, 5489-5492.

81. Yamashita, K., Totani, K., Kuroki, M., Matsuoka, Y., Ueda, I., Kobata, A. (1987) Structural studies of the carbohydrate moieties of carcinoembryonic antigens. Cancer Res., 47, 3451-3459.

82. Hakim, A.A. (1984) Carcinoembryonic antigen, a tumor-associated glycoprotein induces defective lymphocyte function. Neoplasma 31, 385-397.

83. Lloyd, K.O., Old, L.J. (1989) Human monoclonal antibodies to glycolipids and other carbohydrate antigens: dissection of the humoral immune response in cancer patients. Cancer Res., 49, 3445-3451.

84. Young, W.W., Hakomori, S.I., Levine, P. (1979) Characterization of anti-Forssman (anti-Fs) antibodies in human sera: their specificity and possible changes in patients with cancer. J. Immunol., 123, 92-96.

85. Hakomori, S.I. (1985) Aberrant glycosylation in cancer cell membranes as focused on glycolipids: overview and perspectives. Cancer Res., 45, 2405-2414.

86. Hakomori, S.I. (1988) In: Altered glycosilation in tumor cells. Editors: Ch. L. Reading, S.I. Hakomori, D.M. Marcus. New York: Alan R. Liss, p. 207-212.

87. Hakomori, S.I. (1989) Aberrant glycosylation in tumors and tumor-associated carbohydrate antigens. Adv. Cancer Res., 52, 257-331.

88. Abe, K., Hakomori, S.I., Ohshiba, S. (1986) Differential expression of difucosyl type 2 chain (LeY) defined by monoclonal antibody AH6 in different locations of colonic epithelia, various histological types of colonic polyps, and adenocarcinomas. Cancer Res. 46, 2639-2644.

89. Itzkowitz, S.H., Shi, Z.R., Kim, Y.S. (1986) Heterogeneous expression of two oncodevelopmental antigens, CEA and SSEA-1, in colorectal cancer. Histochem. J. 18:155-163.

90. Itzkowitz, S.H., Yuan, M., Gukushi, Y., Palekar, A., Phelps, P.C., Shamsuddin, A.M., Trump, B.F., Hakomori, S.I., Kim, Y.S. (1986) Lewis-x and sialylated Lewis-x-related antigen expression in human malignant and nonmalignant colonic tissues. Cancer Res., 46, 2627-2632.

91. Itzkowitz, S.H., Yuan, M., Ferrell, L.D., Palekar, A., Kim, Y.S. (1986) Cancer-associated alterations of blood group antigen expression in human colorectal polyps. Cancer Res., 46, 5976-5984.

92. Itzkowitz, S.H., Yuan, M., Montgomery, C.K., Kjeldsen, T., Takahashi, H.K., Bigbee, W.L., Kim, Y.S. (1989) Expression of Tn, sialosyl-Tn, and T antigens in human colon cancer. Cancer Res., 49, 197-204.

93. Kim, Y.S., Yuan, M., Itzkowitz, S.H., Sun, Q., Kaizu, T., Palekar, A., Trump, B.F., Hakomori, S.I. (1986) Expression of LeY and extended LeY blood

group-related antigens in human malignant, premalignant, and nonmalignant colonic tissues. Cancer Res., 46, 5985-5992.

94. Lloyd, K.O. (1988) In: Altered glycosylation in tumor cells. Editors: Ch.L. Reading, S.-I. Hakomori; D.M. Markus, NY., Alan R. Liss, pp. 235-243.

95. Stein, R., Chen, S., Grossman, W., Goldenberg, D.M. (1989) Human lung carcinoma monoclonal antibody specific for the Thomsen-Friedenreich antigen. Cancer Res., 49:32-37.

96. Boog, C.J.P., Neefjes, L.J., Boes, J., Ploegh, H.L., Melief, C.J. (1989) Specific immune responses restored by alteration in carbohydrate chains of surface molecules on antigen-presenting cells. Eur. J. Immunol. 19:537-542.

97. Adachi, M., Hayami, M., N. Kashiwagi, N., Mizuta, T., Ohta, Y., Gill, M.J., Matheson, D.S., Tamaoki, T., Shiozawa C., Hakomori, S. (1988) Expression of Ley antigen in human immunodeficiency virus-infected human T cell lines and in peripheral lymphocytes of patients with acquired immune deficiency syndrome (AIDS) and AIDS-related complex (ARC) J. Exp. Med. 167:323-331.

98. Andrews, P.W., Gonczol, E., Fenderson, B.A., Holmes, E.H., O'Malley, G., Hakomori, S., Plotkin, S. (1989) Human cytomegalovirus induces stage-specific embryonic antigen 1 in differentiating human teratocarcinoma cells and fibroblasts. J. Exp. Med. 169:1347-1360.

99. Nichols, E.J., Kannagi, R., Hakomori, S., Krantz, M.J., Fuks, A. (1985) Carbohydrate determinants associated with carcinoembryonic antigen (CEA) J. Immunol. 135:1911-1913.

100. Tanaka, K., T. Yoshioka, C. Bieberich and G. Jay (1988) Role of the major histocompatibility complex class I antigens in tumor growth and metastasis. Annu. Rev. Immunol. 6:359-380.

101. Barbacid, M. (1986) Oncogenes and human cancer: cause or consequence? Carcinogenesis 7:1037-1042.

102. Bos, J.L., Fearon, E.R., Hamilton, S.R., Verlaan-de Vries, M., van Boom, J.H., van der Eb, A.J., Vogelstein, B. (1987) Prevalence of ras gene mutations in human colorectal cancers. Nature 327:293-297.

103. Sakamoto, J., Watanabe, T., Tokumaru, T., Takagi, H., Nakazato, H., Lloyd, K.O. (1989) Expression of Lewis-a, Lewis-b, Lewis-x, Lewis-y, siayl-Lewis-a, and sialyl-Lewis-x blood group antigens in human gastric carcinoma and in normal gastric tissue. Cancer Res. 49:745-752.

104. Igarashi, Y. (personal communication).

105. Kurilsky, P., Claverie, J.M. (1989) MHC restriction, alloreactivity, and thymic education: a common link? Cell, 56:327-329.

106. Buckingham, R.H., Grosjean, H. (1986)NTC: Its control and relevance to living systems. In: Accuracy in Molecular Processes, T.B. Kirkwood, R.F. Rosenberger, D.J. Galas, eds. (London: Chapman & Hall) pp. 83-126.

107. Manley, J. L. (1978).Synthesis and degradation of termination and premature-termination fragments

of beta-galactosidase in vitro and in vivo. J. Mol. Biol. 125:407-432.

108. Feizi, T. (1985) Demonstration by monoclonal antibodies that carbohydrate structures of glycoproteins and glycolipids are onco-developmental antigens. Nature. 314:53-57.

109. Hakomori, S.I., Kannagi, R. (1983) J. Natl. Cancer Inst. 71:231-251.

110. Miyake, M., Zenita, K., Tanaka, O., Okada, Y., Kannagi, R. (1988) Stage-specific expression of SSEA-1-related antigens in the developing lung of human embryos and its relation to the distribution of these antigens in lung cancers. Cancer Res. 48:7150-7158.

111. Solter, D., Knowles, B.B. (1978) Monoclonal antibody defining a stage-specific mouse embryonic antigen (SSEA-1) Proc. Natl. Acad. Sci. USA 75:5565-5569.

112. Gooi, H.C., Feizi, T., Kapadia, A., Knowles, B.B., Solter, D., Evans, M.J. (1981) Stage-specific embryonic antigen involves alpha 1 goes to 3 fucosylated type 2 blood group chains. Nature. 292:156-158.

113. Kannagi, R., Nudelman, E., Levery, S.B., Hakomori, S. (1982) A series of human erythrocyte glycosphingolipids reacting to the monoclonal antibody directed to a developmentally regulated antigen SSEA-1. J. Biol. Chem. 257:14865-14874.

114. Abe, K., J.M. McKibbin and S.I. Hakomori. (1983) The monoclonal antibody directed to difucosylated type 2 chain. J. Biol. Chem. 258:11793-11797.

115. Fukushi, Y., Kannagi, R., Hakomori, S., Shepard, T., Kulander, B.G., Singer, J.W. (1985) Location and distribution of difucoganglioside in normal and tumor tissues defined by its monoclonal antibody FH6. Cancer Res. 45:3711-3717.

116. Fukushi, Y., Hakomori, S., Nudelman, E., Cochran, N. (1984) Novel fucolipids accumulating in human adenocarcinoma. II. Selective isolation of hybridoma antibodies that differentially recognize mono-, di- and trifucosylated type 2 chain. J. Biol.Chem. 259:4681-4685.

117. Kannagi, R., Fukushi, Y., Tachikawa, T., Noda, A., Shin, S., Shigeta, K., Hiraiwa, N., Fukuda, Y., Inamoto, T., Hakomori, S., Imura, H. (1986) Quantitative and qualitative characterization of human cancer-associated serum glycoprotein antigens expressing fucosyl or sialyl-fucosyl type 2 chain polylactosamine. Cancer Res. 46:2619-2626.

118. Shi, Z.R., McIntyre, L.J., Knowles, B.B., Solter, D., Kim, Y.S. (1984) Expression of a carbohydrate differentiation antigen, stage-specific embryonic antigen 1, in human colonic adenocarcinoma. Cancer Res. 44:1142-1147.

119. Spitalnik, S.L., Spitalnik, P.F., Dubois, C., Mulshine, J., Magnani, J.L. Cuttitta, F., Civin, C.I., Minna, J.D., Ginsburg, V. (1986) Glycolipid antigen expression in human lung cancer. Cancer Res. 46:4751-4755.

120. Cuttita, F., Rosen, S., Gazdar, A.F., Minna, J.D. (1981) Proc. Natl. Acad. Sci. USA 78:4591-4595.

121. Huang, L.C., Brockhaus, M., Magnani, J.L., Cuttitta, S.R., J.D., Ginsburg, V. (1983) Many mono-

clonal antibodies with an apparent specificity for certain lung cancers are directed against a sugar sequence found in lacto-N-fucopentaose III. Arch. Biochem. Biophys. 220:318.

122. Chia, D., Terasaki, P.I., Suyama, N., Galton, J., Hirota, M., Katz, D. (1985) Use of monoclonal antibodies to sialylated Lewis-x and sialylated Lewis-a for serological tests of cancer. Cancer Res. 45:435-437.

123. Zenita, K., Kirihata, Y., Kitahara, A., Shigeta, K., Higuchi, K., Hirashima, K., Murachi, T., Miyake, M., Takeda, T., Kannagi, R. (1988) Fucosylated type-2 chain polylactosamine antigens in human lung cancer. Int. J. Cancer 41:344-349.

124. Andrews, P.W., Goodfellow, P.N., Shevinsky, L.H., Bronson, D.L., Knowles, B.B. (1982) Cell-surface antigens of a clonal human embryonal carcinoma cell line: morphological and antigenic differentiation in culture. Int. J. Cancer. 29:523-531.

125. Combs, S.G., Marder, R.J., Minna, J.D., Mulshine, J.L., Polovina, M.R., Rosen, S.T. (1984) Immuno-histochemical localization of the immunodominant differentiation antigen lacto-N-fucopentaose III in normal adult and fetal tissues. J. Histochem. Cytochem. 32:982-988.

126. Fox, N., Damjanov, I., Martinez-Hernandez, A., Knowles, B.B., Solter, D. (1981) Immunohisto-chemical localization of the early embryonic antigen (SSEA-1) in postimplantation mouse embryos and fetal and adult tissues. Dev. Biol. 83:391-398.

127. Fox, N., Damjanov, I., Knowles, B.B., Solter, D. (1983) Immunohistochemical localization of the mouse stage-specific embryonic antigen 1 in human tissues and tumors. Cancer Res. 43:669-678.

128. Yamamoto, M., Boyer, A.M., Schwarting, G.A. (1985) Fucose-containing glycolipids are stage- and region-specific antigens in developing embryonic brain of rodents. Proc. Natl. Acad. Sci. USA 82:3045-3049.

129. Springer, G.F., Desai, P.R. (1977) Cross-reacting carcinoma-associated antigens with blood group and precursor specificities. Transplant. Proc. 9, 1105.

130. Springer, G.F., Desai, P.R, Banatwala, I. (1975) Blood group MN antigens and precursors in normal and malignant human breast glandular tissue. J. Natl. Cancer Inst. 54, 335.

131. Howard, D.R., Taylor, C.R. (1980) An antitumor antibody in normal human serum: Reaction of anti-T with breast carcinoma cells. Oncology 37, 142.

132. Boland, C.R., Montgomery, C.K., Kim, Y.S. (1982) Alterations in human colonic mucin occurring with cellular differentiation and malignant trans-formation.Proc. Natl. Acad. Sci. USA, 79, 2051

133. Orntoft, T.F., Mors, N.O., Eriksen, G., Jacobsen, N.P., Poulsen, H.S. (1985) Comparative immuno-peroxidase demonstration of T-antigens in human colorectal carcinomas and morphologically abnor-mal mucosa. Cancer Res., 45:447-452.

134. Anglin, J.H., Lerner, M.P., Nordquist, R.E. (1977) Blood group-like activity released by human mam-mary carcinoma cells in culture. Nature (London)

269, 254.

135. Laurent, J.C., Noel, P., Faucon, M. (1978) Expres-sion of a cryptic cell surface antigen in primary cell cultures from human breast cancer [letter] Biomedi-cine 29, 260.

136. Limas, C., Lange, P. (1980) Altered reactivity for A, B, H antigens in transitional cell carcinomas of the urinary bladder. A study of the mechanisms in-volved. Cancer 46, 1366.

137. Coon, J., Weinstein, R.S., Summers, J. (1982) Blood group precursor T-antigen expression in human urinary bladder carcinoma. Am. J. Clin. Pathol. 77, 692.

138. Summers, J.L. et al. (1983) Prognosis in carcinoma of the urinary bladder based upon tissue blood group abh and Thomsen-Friedenreich antigen status and karyotype of the initial tumor. Cancer Res. 43, 934.

139. Lehman, T.P., Cooper, H.S., Mulholland, S.G. (1984) Peanut lectin binding sites in transitional cell carci-noma of the urinary bladder. Cancer 53, 272.

140. Ghazizadeh, M., S. Kagawa, S., Izumi, K., Kurokawa, K. (1984) Immunohistochemical localization of T antigen-like substance in benign hyperplasia and adenocarcinoma of the prostate. J. Urol.,132:1127-1130.

141. Springer, G.F., Murthy, M.S., Desai, P.R., Scanlon, E.F. (1980) Breast cancer patient's cell-mediated immune response to Thomsen-Friedenreich (T) anti-gen Cancer 45, 2949.

142. Springer, G.F., Ansell, N.J. (1958) Proc. Natl. Acad. Sci. USA, 44, 182.

143. Lloyd, K.O., Kabat, E.A. (1968) Immunochemical studies on blood groups. XLI. Proposed structures for the carbohydrate portions of blood group A, B, H, Lewis-a, and Lewis-b substances. Proc. Natl. Acad. Sci USA, 61, 1470.

144. Friedenreich, V. (1930) The Thomsen Hemaggluti-nation Phenomenon (Levin & Munksgaard, Copenhagen)

145. Burnet , F.M., Anderson, S.G. (1947) J. Exp. Biol. Med., 25, 213.

146. Caselitz, F.H., Stein, G. (1953) Immunitaetsforsch. 110, 165.

147. Dausset, J., Moullec, Bernard, J. (1959) Blood 14, 1079.

148. Springer, G.F., Tegtmeyer, H., Cromer, D.W. (1984) Fed. Proc. Fed. Am. Soc. Exp. Biol., 43, 6.

149. Springer, G. F., P.R., Desai, P.R., Murthy, M.S., Tegtmeyer, H., Scanlon, E.F. (1979) Human carci-noma-associated precursor antigens of the blood group MN system and the host's immune responses to them. In: Progress in Allergy, Kallos, P., Waksman, B.H., deWeck, A.L., Ishizaka, K. Eds. (Karger, Basel), vol. 29, pp. 42-96.

150. Springer, G.F., Desai, P.R. (1982) Detection of lung- and breast carcinoma by quantitating serum anti-T IgM levels with a sensitive, solid-phase immunoas-say. Naturwissenschaften. 69, 346.

151. Desai, P., Springer, G. (1984) In: Protides of the Biological Fluids, H. Peeters, Ed. (Pergamon, Ox-ford), vol. 31, pp. 421-424.

152. Springer, G.F. , Taylor, C.R., Howard, D.R., Tegtmeyer, H., Desai, P.R., Murthy, S.M., Felder, B., Scanlon, E.F. (1985) Tn, a carcinoma-associated antigen, reacts with anti-Tn of normal human sera. Cancer, 55:561-569.

153. Springer, G.F. et al. (1983) New approaches in biology, diagnosis, and treatment. In: Cellular Oncology, P.J. Moloy and G.L. Nicolson, Eds. (Praeger, New York), vol. 1, pp. 99-130.

154. Goodale, R.L., Springer, G.F., Shearen, J.G., Desai, P.R., Tegtmeyer, H. (1983) Delayed-type cutaneous hypersensitivity to Thomsen-Friedenreich (T) antigen in patients with pancreatic cancer. J. Surg. Res. 35, 293.

155. Springer, G.F., Desai, P.R. (1974) Common precursors of human blood group MN specificities. Biochem. Biophys. Res. Commun. 61, 470.

156. Kim, Z., Uhlenbruck, G. (1966) Studies on the T-antigen and T-agglutinin. Immunitaetsforsch. Exp. Ther. 130, 88.

157. Springer, G.F., Desai, P.R. (1975) Human blood-group MN and precursor specificities: structural and biological aspects. Carbohydr. Res. 40, 183.

158. Bray, J., Lemieux, R.U., McPherson, T.A. (1981) Use of a synthetic hapten in the demonstration of the Thomsen-Friedenreich (T) antigen on neuraminidase-treated human red blood cells and lymphocytes. J. Immunol. 126, 1966.

159. Springer, G.F., Desai, P.R., Schachter, H., Narasimhan, S. (1976) Enzymatic synthesis of human blood group M-, N- and T-specific structures.Naturwissenschaften. 63, 488 .

160. Carton, J.P. et al. (1978) Eur. J. Biochem. 92, 111.

161. Desai, P.R., Springer, G.F. (1979) Biosynthesis of human blood group T-, N- and M-specific immunodeterminants on human erythrocyte antigens. J. Immunogenet. 6, 403.

162. Hirohashi, S., Clausen, H., Yamada, T., Shimosato, Y., Hakomori, S. (1985) Blood group A cross-reacting epitope defined by monoclonal antibodies NCC-LU-35 and -81 expressed in cancer of blood group O or B individuals: Its identification as Tn antigen. Proc. Natl. Acad. Sci. USA, 82, 7039-43.

163. Takahashi, H., Metoki, R., Hakomori, S. (1988) Immunoglobulin G3 monoclonal antibody directed to Tn antigen (tumor-associated alpha-N-acetylgalactosaminyl epitope) that does not cross-react with blood group A antigen.Cancer Res., 48, 4361-67.

164. Kurosaka, A., Kitagawa, H., S. Fukui, S., Numata, Y., Nakada, H., Funakoshi, I., Kawasaki, T., Ogawa, T., Iijima, H., Yamashina, I. (1988) A monoclonal antibody that recognizes a cluster of a disaccharide, NeuAc alpha(2-6)GalNAc, in mucin-type glycoproteins. J. Biol. Chem., 263, 8724-26.

165. Kjelden, T., Clausen, H., Hirohashi, S., Ogawa, T., Iijima H., Hakomori, S. (1988) Preparation and characterization of monoclonal antibodies directed to the tumor-associated O-linked sialosyl-2-6 alpha-N-acetylgalactosaminyl (sialosyl-Tn) epitope. Cancer Res., 48, 2214-20.

166. Johnson, V.G., Schlom, J., Paterson, A.J., Bennett, J., Magnani, J.H., Colcher, D. (1986) Analysis of a human tumor-associated glycoprotein (TAG-72) identified by monoclonal antibody B72.3. Cancer Res., 46, 850-57.

167. Longenecker, B.M., MacLean, G.D., McEwan, A.J., Sykes, T., Henningsson, C., Suresh, M.R., Noujaim, A.A. (1988) In: Altered Glycosylation in tumor cells. Editors: Ch. L. Reading, S.I. Hakomori, D.M. Markus, N.Y., Eds. Alan R. Liss, pp. 307-320.

168. Galili, U., Rachmilewitz, E.A., Peleg, A., Flechner, I. (1984) A unique natural human IgG antibody with anti-alpha-galactosyl specificity. J. Exp. Med. 160:1519-1531.

169. Galili, U., Macher, B.A., Buehler, J., et al. (1985) Human natural anti-alpha-galactosyl IgG. II. The specific recognition of alpha (1-3)-linked galactose residues. J. Exp. Med. 162:573-582.

170. Galili, U., Buehler, J., Shohet, S.B., et al. (1987) The human natural anti-Gal IgG. III. The subtlety of immune tolerance in man as demonstrated by crossreactivity between natural anti-Gal and anti-B antibodies. J. Exp. Med. 165:693-704.

171. Galili, U., Mandrell, R.E., Hamadeh, R.M. et al. (1988) Interaction between human natural anti-alpha-galactosyl immunoglobulin G and bacteria of the human flora. Infect. Immun. 56:1730-1737.

172. Galili, U., Shohet, S.B., Kobrin, E., et al. (1988) Man, apes, and Old World monkeys differ from other mammals in the expression of alpha-galactosyl epitopes on nucleated cells. J. Biol. Chem. 263:1755-1762.

173. Galili, U., Clark, M.R., Shohet, S.B., et al. (1987) Evolutionary relationship between the natural anti-Gal antibody and the Gal alpha 1-3Gal epitope in primates. Proc. Natl. Acad. Sci USA 84:1369-1373.

174. McCoy, J.P., Goldstein, I.G., Varani, J. (1985) Tumor Biology 6:99-114.

175. Varani, J., Lovet, E.J. III, Wicha, M., et al. (1983) Cell surface alpha-D-galactopyranosyl end groups: use as markers in the isolation of murine tumor cell lines with different cancer-causing potentials. J. Natl. Cancer Inst. 71:1281-1287.

176. Grimstad, I.A., Varani, J., McCoy, J.P. (1984) Contribution of alpha-D-galactopyranosyl end groups to attachment of highly and low metastatic murine fibrosarcoma cells to various substrates. Exp. Cell Res. 155:345-358.

177. Grimstad, I.A., Bosnes, V. (1987) Cell-surface laminin-like molecules and alpha-D-galactopyranosyl end-groups of cloned strongly and weakly metastatic murine fibrosarcoma cells. Int. J. Cancer, 40:505-510.

178. Castronovo, V., Colin, C., Parent, B., Foidart, J.M., Lambotte, R., Mahieu, Ph. (1989) Possible role of human natural anti-Gal antibodies in the natural antitumor defense system. J. Natl. Cancer Inst. 81:212-216.

179. Galili, U., Korkesh, A., Kahana, I., Rachmilewitz, E.A. (1983) Demonstration of a natural antigalactosyl IgG antibody on thalassemic red blood cells. Blood,

61:1258-1264.

180. Galili, U., Clark, M.R., Shohet, S.B. (1986) Excessive binding of natural anti-alpha-galactosyl immunoglobin G to sickle erythrocytes may contribute to extravascular cell destruction. J. Clin. Invest., 77:27-33.

181. Elliott, B.E., Carlow, D.A., Rodricks, A.-M., Wade, A. (1989) Perspectives on the role of MHC antigens in normal and malignant cell development. Adv. Cancer Res., 53:181-245.

182. Ottesen, S.S., Fromholt, V., Kieler, J. (1988) Correlation between classification of human urothelial cell lines and HLA-A,B,C expression. Cancer Immunol. Immunother., 26:83-86.

183. Codinton, J.F. (1975) Masking of cell-surface antigens on cancer cells. In Cellular membranes and tumor cell behavior. Williams and Wilkins Co., Baltimore, MD. p.399.

184. Chandrasekaran, E.V., Davila, M., Nixon, D.W., Goldfarb, M., Mendicino, J. (1983) Isolation ans structures of the oligosaccharide units of carcinoembryonic antigen. J. Biol. Chem., 258:7213-7222.

185. Marz, L., Bahl, O.P. (1973) Blood-group activity of human chorionic gonadotropin. Biochem. Biophys. Res. Commun., 55:717-723.

186. Kim, U., Baumler, A., Carruthers, C., Bielat, K. (1975) Immunological escape mechanism in spontaneously metastasizing mammary tumors. Proc. Natl. Acad. Sci. USA, 72:1012-1016.

187. Kim, U. (1970) Metastasizing mammary carcinomas in rats: induction and study of their immunogenicity. Science, 164:72-74.

188. Alexander, P. (1974) Proceedings: Escape from immune destruction by the host through shedding of surface antigens: is this a characteristic shared by malignant and embryonic cells? Cancer Res., 34:2077-2082.

189. Currie, G.A., Alexander, P. (1974) Spontaneous shedding of TSTA by viable sarcoma cells: Its possible role in facilitating metastatic spread. Br. J. Cancer, 29:72-75.

190. Pellis, N.R., Kahan, B.D. (1975) Specific tumor immunity induced with soluble materials: restricted range of antigen dose and of challenge tumor load for immunoprotection. J. Immunol., 115:1717-1722.

191. Fung, P.Y.S., Madej, M., Koganty, R.R., Longenecker, B.M (1990) Active specific immunotherapy of a murine mammary adenocarcinoma using a synthetic tumor-associated glycoconjugate. Cancer Res., 50:4308-4314.

192. Singhal, A., Fohn, M., Hakomori, S. (1991) Induction of α-N-acetylgalactosamine-O-Serine/Threonine (Tn) antigen-mediated cellular immune response for active immunotherapy in mice. Cancer Res., 51:1406-1411.

193. Miller, J.F.A.P. (1989) Tolerance and the thymus. Transplantation Proceeding, 21:59-60.

194. Barnd, D.L., Lan, M.S., Metzgar, R.S., Finn, O.J. (1989) Specific, major histocompatibility complex-unrestricted recognition of tumor-associated mucins by human cytotoxic T cells. Proc. Natl. Acad. Sci. USA, 86:7159-7163.

195. Xing, P.X., Tjandra, J.J., Reynolds, K., McLaughlin, P.J., Purcell, D.F.J., McKenzie, I.F.C. (1989) Reactivity of anti-human milk fat globule antibodies with synthetic peptides. J. Immunol., 142:3503-3509.

196. Metzgar, R.S., Rodriguez, N., Finn, O.J., Lan, M.S., Daasch, V.N., Fernsten, P.D., Meyers, W.C., Sindelar, W.F., Sandler, R.S., Seigler, H.F. (1984) Detection of a pancreatic cancer-associated antigen (DU-PAN-2 antigen) in serum and ascites of patients with adenocarcinoma. Proc. Natl. Acad. Sci. USA, 81:5242-5246.

197. Bossman, H.B., Hall, T.C. (1974) Enzyme activity in invasive tumors of human breast and colon. Proc. Natl. Acad. Sci. USA, 71:1833-1837.

198. Bossman, H.B., Bieber, G.F., Brown, A.E., Case, K.R., Gersten, D.M., Kimmerer, T.W., Lione, A. (1973) Biochemical parameters correlated with tumour cell implantation. Nature, 246:487-489.

199. Dobrossy, L., Pavelic, Z.P., Bernacki, R.J. (1981) A correlation between cell surface sialyltransferase, sialic acid, and glycosidase activities and the implantability of B16 murine melanoma. Cancer Res., 41:2262-2266.

200. Lampson, L.A., Fisher, C.A., Whelan, J.P. (1983) Striking paucity of HLA-A, B, C and β_2-microglobulin on human neuroblastoma cell lines. J. Immunol., 130:2471-2474.

CHAPTER 6

THE BGA-RELATED GLYCOEPITOPES AND CANCER METASTASIS

Invasion and metastasis of cancer cells are complex multistage processes affected by a multiplicity of mechanisms and involve numerous factors.[1-3] The ability of tumor cells to detach from the primary lesion, to penetrate basement membrane surrounding blood vessels, and finally to implant into remote tissues and organs is mediated in part by hydrolytic enzymes.[4,5] The spectrum of biological activities of the plasminogen activation system makes it an ideal candidate to play a key role in cancer invasion[4,6] and there are numerous data supporting the hypothesis concerning the role of plasminogen activation system in tumor invasion and metastasis.[7-13] Furthermore, the aberrant adhesion behavior of cancer cells contributes significantly to the metastatic dissemination. The adhesion of transformed cells is characterized by preservation of cellular recognition functions mediated by carbohydrate and peptide cell adhesion epitopes. Inability to display the secondary stable attachment and strong terminal tissue-specific adhesion, impairment of adequate intracellular signaling and biological cellular response are also typical for cancer cell adhesive behavior. Carbohydrate-mediated cell-cell and cell-substratum adhesion play a critically important role in cancer metastasis. Even for invasion the attachment of cancer cells to the basement membrane and extracellular matrix is required. Cancer cell adhesion is critically important for arrest of circulating tumor cells at the distant specific site(s) of metastasis.

Blood capillaries may act as a sieve and thus arrest tumor cells with a diameter bigger than that of blood capillaries. However, tumor cells change their shape to pass through narrow capillaries.[134] Lymph node transmigration has been tested by intralymphatic injection of tumor cells. Fisher and Fisher[135] found that different types of carcinoma cells rapidly moved through lymph nodes. Kurokawa[136] was observed that hepatoma cells were retained in the regional lymph nodes and he did not detect tumor cells elsewhere until macroscopic metastases had developed in the regional nodes. Therefore, tumor cells are retained in target organs because they adhere to the other cells and/or to the endothelium of the organ's capillaries.

CANCER CELL AGGREGATION AND METASTASIS

One of the characteristic features of the growth of malignant tumors cultivated in vitro is their ability to form three-dimensional, spheroid-like structures termed multicell spheroids.[14,15] Similar spheroid-like formations are observed in rodent and human malignant tumors in vivo, according to biochemical, immunocytochemical, morphological and cytological criteria. Therefore, multicell spheroids in vitro are considered a cytofunctional analog of the avascular stage of in vivo tumor development.[15] Cell association and aggregation occur in the first stage of multicell spheroid formation and can be measured in vitro using the cell-aggregation assay. Many studies indicate that cell surface components play an important role in metastasis,[16-18] in particular in formation of tumor cell aggregates and emboli. Circulating metastatic tumor cells tend to adhere to other tumor cells (homotypic adhesion) or to host cells including platelets and lymphocytes (heterotypic adhesion) to form emboli, the formation of which is correlated with metastasis.[19-21] The lodgment, attachment, and growth of blood-borne neoplastic cells depend largely on embolization.[22,23] In a B16 melanoma experimental model of metastasis,[24,25] it has been shown that tumor cell clumps produce more lung metastases after IV injection than do single cells[19] and a correlation has been demonstrated between the tendency of cells to undergo homotypic and heterotypic aggregation in vitro and their metastatic potential in vivo.[23,26-29] On the other hand, among hepatomas, low homotypic adhesion correlated with high metastatic capacity,[92] and it has been suggested that decrease in homotypic adhesion possibly causes the cancer cell release and dissemination from the primary tumor.[18]

Most of these studies have demonstrated differences in cell surface glycoconjugates between tumor cells exhibiting high and low metastatic capacity. It is quite well established that cell surface constituents are involved in cell-cell and cell-substratum recognition and adhesion[30] and that such interactions are relevant for tumor cell embolization, specific arrest, and implantation in secondary sites.[16-18] The emboli may lodge nonspecifically in narrow capillaries, or they may adhere preferentially in specific organs, presumably as a result of cognitive interactions between the tumor cells and capillary endothelial cells or exposed basement membrane components.[16,18] Tumor cell surface lectins play major roles in cellular

interactions in vivo that are important for the formation of emboli and for the arrest of circulating tumor cells.[31-35] Antilectin antibodies localize lectins in the cytoplasmic compartment of various normal cells as well as in the extracellular compartment, where they seem to be associated with the extracellular matrix after being secreted by the cells.[36-40]

Various functions have been proposed for the lectins, including mediation of intercellular recognition and adhesion,[40-46] control of cellular proliferation,[36,47,48] organization of the extracellular matrix[36,38,40] and elastic fiber formation.[49] Tumor cells and tissues contain lectins that are similar to those isolated from normal cells in their sugar-binding specificities, molecular weights, and sequences.[35,50-58] It was found by using monoclonal and polyclonal antilectin antibodies that the lectins were expressed on the cell surface and that their level correlated with malignant transformation and metastatic propensity.[54,55,57] A key role has been suggested for the tumor-surface lectins in anchorage-independent growth in vitro and in tumor embolization (homotypic and heterotypic aggregation) during the metastatic process in vivo.[31-35] A number of rodent and human malignant tumors express lectins with biospecificity to galactose, lactose and N-acetyl-lactosamine.[45,52,59,60] It has been considered, and confirmed for quite some time, that adhesion of platelets to tumor cells is a prerequisite for cancer metastasis.[61-65]

The homotypic aggregation of B16 melanoma cells is dependent on the presence of fetal bovine serum.[23,29] One possible explanation for this serum requirement could be that a serum glycomacromolecule(s) mediates intercellular adhesion similar to the action of cell-cell adhesion molecules. Our studies have shown that the serum glycoamines and other macromolecules may play important roles in changing the adhesion and aggregation properties of cancer cells in vitro and metastasis in vivo and thus may be involved in the process of modification of cell recognition, association and aggregation.[66-70] Since carbohydrates and carbohydrate-recognition structures are involved in cell aggregation, we have initiated an investigation of structural-functional interrelations of glycoamines, their synthetic structural analogs, and high molecular weight carbohydrate-associated serum proteins as modifiers (inhibitor or stimulator) in vitro of tumor cell aggregation, and metastasis in vivo. The results of ongoing research programs [69,70] have provided increasing evidence that support the hypothesis that: a) different molecular isoforms of glycomacromolecules in serum contain cell-originated glycoepitopes which are involved in cell recognition and cell association processes (the family of cell recognition glycoepitopes containing extracellular biopolymers [biomacromolecules]—CRG-CEB); b) these different molecular isoforms of the CRG-CEB and serum carbohydrate-binding proteins, e.g., anticarbohydrate antibodies, lectins, etc., play opposite roles, either as inhibitors and/or stimulators of cell association and aggregation, depending on the nature and size of the carrier molecules, valence of glycoepitopes (for CRG-CEB) and binding properties (for carbohydrate-binding proteins); c) human cancer is accompanied by changes in quantitative and qualitative characteristics of the CRG-CEB and carbohydrate-binding proteins in blood serum; and d) benign and malignant human tumors have different in vitro cell aggregation properties.

Thus, the homotypic and heterotypic aggregation properties of tumor cells represent important biological features of malignancy because these properties of transformed cells could determine the metastatic ability of neoplastic tumors.[16-23,26,27,29,31-35,49] The key structural determinants of the tumor cells that participate in cell recognition, association and aggregation have been determined as carbohydrates and carbohydrate-binding proteins.[16-18,30-36,45] Generally, tumor cells metastasize to the organ containing the first capillary bed encountered after leaving the primary tumor; however, other tumors exhibit preferential metastasis to specific organs: ocular melanoma metastasizes to the liver, and clear cell carcinoma of the kidney spreads to the thyroid.[16,71] It is clear that not only tumor cells play an exclusive role in metastatic spread, but some other cellular counterparts, e.g., leukocytes, platelets, are directly involved in metastasis. There is a problem in obtaining metastasis by injecting human tumor cell lines in nude mice: most do not metastasize in nude mice, even if originally derived from a metastatic lesion.[72] It is possible that some of the important host cellular metastatic partners are missing in this experimental system.

THE BGA-RELATED GLYCOEPITOPES AS KEY STRUCTURAL DETERMINANTS OF CANCER CELL AGGREGATION AND ADHESION

Aberrant glycosylation of cell membrane macromolecules is one of the universal attributes of malignant tumors. A rather limited number of molecular probes based on monoclonal anticarbohydrate antibodies now enables the detection of over 90% of the most widespread forms of human cancer.[73,74] The most characteristic manifestation of aberrant glycosylation of cancer cells is neosynthesis (or ectopic synthesis), the synthesis of incompatible antigens, and incomplete synthesis (with or without the accumulation of precursors) of the BGA-related glycoepitopes.[73-75] As we have mentioned earlier, BGA-related glycoepitopes are directly involved in the homotypic (tumor cells, embryonal cells) and heterotypic (tumor cells-normal cells) formation of cellular aggregates, e.g., Lewis X antigens, polylactosamine sequences and T- and Tn-antigens, which have been demonstrated in different experimental systems.[76-80] BGA-related alterations in tissue glycosylation patterns are detected in benign (premalignant) tumors with high risk of malignant transformation, in primary malignant tumors, and in metastases,[81-86] i.e., they were demonstrated as typical alterations in different stages of tumor progression.

Recently, experimental evidence indicated that some of the BGA-related glycodeterminants, which had been identified earlier as tumor-associated carbohydrate antigens (TACA), function as key adhesion molecules.[70,87,88] Furthermore, the expression of some TACA (sialosyl-LeX; sialosyl-Tn; H/LeY/LeB) distinguishes metastatic deposits from the primary tumor and indicates markedly poorer prognosis in cancer patients with colonic, ovarian, and lung cancer.[87,89,90] It has been suggested that tumor cell population expressing particular carbohydrate antigens (sialosyl-Tn, sialosyl-LeX; H; or LeY) may preferentially initiate invasion and metastatic spread by an unknown mechanism.[87] Earlier the same conclusion has been considered relative to the other carcinoma-associated carbohydrate antigens: T- and Tn- antigens[77,78] and Galα (1-3) Gal epitope.[91] The common cell-biological property of those glycoantigens is their function as key adhesion molecules that initiate a latent multistep "cascade" cell adhesion mechanism, and those glycoantigens are responsible for homotypic and heterotypic cancer cell aggregation.

A Model of Tissue and/or Organ Specificity of Cancer Metastasis

Metastases are often distributed in a manner not simply deducible from the anatomic position in the body of the primary tumor. There are characteristic metastatic distribution patterns for many types of human neoplasm.[93] In general, dissemination of cancer cells is a multistep or "cascade" process when the tumor cells first seed to one or a few sites that are responsible for the subsequent production of larger numbers of peripheral metastases in different organs.[93-96] For solid tumors, the typical "key" or "generalizing" organs are lungs, liver and bones, whereas hematologic malignancies (myeloma and leukemias) usually disseminate via the spleen or liver.[18] Human tumors often metastasize in an unexpected manner:[97] there is often a preference for one single organ, e.g., adrenals in the case of lung carcinoma, or there are associations between organs, e.g., simultaneous metastasis in brain and adrenals is remarkably common.

Two concepts explaining the observed organ distributions of cancer metastases have been put forward:[98] "seed and soil" hypothesis emphasizing some kind of affinity between metastatic tumor cells and specific organs, and the concept emphasizing vascular connections and anatomical relationships between primary tumor and the sites of its metastases. Ewing[99] hypothesized that metastases occur as a result of purely mechanical factors which

are due to the anatomic structure of the vascular system. The "seed and soil" hypothesis envisions that the microenvironment of individual organs may allow for specific implantation, survival and growth of tumor cells with unique properties.[100] These concepts have been considered together or separately in order to explain the secondary (metastatic) site preference of certain neoplasm.[16,18,101-105]

Numerous clinical studies show clearly the secondary site preference for metastatic colonization for certain human cancer. Clear cell carcinoma of the kidney is one of the most frequent types of metastasis found in thyroid gland;[106] prostatic carcinoma metastasizes to bone at very high frequencies;[107] small cell carcinoma of the lung often spreads to the brain;[108] neuroblastoma preferentially colonizes liver and adrenal glands.[16,109] The regional metastatic involvements could be explained for the most part by anatomical or mechanical factors such as efferent venous circulation or lymphatic drainage to regional lymph nodes.[105] However, this generalization does not explain distant organ colonization by metastatic tumor cells. In many cases different types of human cancer establish their own patterns of metastatic dissemination.[16,105]

Experiments with transplanted organ grafts as target sites for distant blood-borne colonization[110-113] strongly suggest that unique specific tumor cell colonization pathways may exist and that determines organ preference of metastasis. Nonetheless, the lung, the first organ encountered by circulating metastatic cells released from the site of primary tumor, is usually the most frequent site of distant metastatic colonization, suggesting that metastases often develop due to passive mechanical trapping of blood-borne tumor emboli.[16] By injection of Walker 256 carcinoma cells into the circulation at different entry points, Griffiths and Salsburg[114] showed that the first organ encountered is probably the most important factor in metastatic colonization. Intraportal vein injection of Walker 256 carcinoma cells caused exclusive formation of liver metastases, while tail vein injection produced only lung tumor colonies.[114] Roos and Dingemans[115] injected the lung-selected murine melanoma subline B16-F10 into the portal veins of mice and found that widespread metastatic colonization of the liver occurred, in contrast to the lung colonization in the experiments in which B16-F10 melanoma cells were injected via the tail vein[103,112,116-121] or left ventricle of the heart.[119,122] Therefore, the secondary site preference of metastatic cell colonization could be determined by the pathway of cancer cell dissemination from the primary tumor site.

Tumor cells may invade, detach and enter the lymph to rest subsequently in near or distant lymph nodes or invade the blood circulatory system where they can lodge in the microcirculation at distant organs. Malignant neoplasms that invade the blood circulatory system should be more likely to develop distant metastases in the lung, the first organ microcirculatory system encountered, while malignancies in the viscera should spread to other abdominal sites through regional lymphatics and local blood circulatory systems for similar reasons.[1,105,123] Proctor[124] has suggested that the initial venous spread of cancer cells is mainly influenced by mechanical factors, whereas the course of the subsequent arterial spread is mainly determined by "soil" effects.

Many patients with malignant melanoma with their first evidence of metastatic disease in major organs had metastases at only one distant site, predominantly lung, followed by liver and then brain.[125,126] However, as the cancer progressed, other distant organ became involved (gastrointestinal tract, bone, etc.). There are classic examples where circulating tumor cells initially lodge in one organ, then detach, recirculate, and find their way to another organ(s) for secondary metastatic colonization.[1,102,103,119-121,127] Tumor cells can apparently pass freely from lymph to blood[127-128] and, therefore, lymphatic and blood-borne models of metastatic dissemination are not mutually exclusive.[16]

How can we consider the general mechanism of cancer cell metastasis, from the point of view of current concepts of vascular cell biology, leukocyte-endothelium interaction, and tumor cell biology? It seems at least questionable that cancer cells may have a definite specific structural determinant(s) that is responsible for their targeting of a specific site(s) of metastasis. There is no single specific structure(s) that determines specificity of leukocyte-endothelial cell recognition and adhesion. There is a multistep, sequential, dynamic "cascade-like" process that determines the unique specificity of the molecular mechanism of leukocyte-endothelial cell recognition, stable attachment, and leukocyte extravasation. This complex mechanism needs a unique molecular machinery of receptor-counterreceptor pairs to be expressed and disappear sequentially at the surface of interacting cells that seems to be a result of advanced differentiation program of those particular type of cells. It is very unlikely that cancer cells could preserve this unique specific differentiation function, since dedifferentiation and loss of features of advanced differentiation are basic fundamental attributes of malignancy. However, it is very important that in many (if not all) types of cancer, the cancer cells retain to a great extent their ability to reveal the initial carbohydrate-mediated stage of cell adhesion: at the same time cancer is characterized by profound disturbances in the subsequent stages of cell adhesion occurring with the involvement of extracellular matrix proteins and integrins, and completed with the formation of specific "gap junctions."

It has been considered that some of the tumor-associated carbohydrate antigens, particularly BGA-related glycoantigens, may play key roles in homotypic and heterotypic cancer cell recognition, aggregation, and adhesion.[70,87,88] The involvement of T- and Tn-antigens (general carcinoma-associated carbohydrate antigens) in liver and brain cancer cell metastasis has previously been considered,[77,78] and the participation of $Gal\alpha$ (1-3) Gal epitopes in cancer cell adhesion to extracellular matrix proteins has been shown.[91] The possibilities of adhesion of tumor cells to human lung capillary endothelia via interaction of H-LeY glycoantigens, and cancer cells to leukocytes and platelets as well as cancer cells to endothelia via selectins-sialylLeX/sialyl LeA/LeX interactions have recently been discussed.[87,88,129] Identification of H/LeY/LeB glycoantigens as determinants of cancer cell motility (migration) extend the number of those examples.[130]

It is clear, however, from the results of recent studies of vascular cell biology and leukocyte-endothelial cell recognition and adhesion that those initial early carbohydrate-mediated events themselves do not provide a basis for specificity of endothelial cell-circulating cell recognition, stable attachment, adhesion and subsequent extravasation of circulating cells. This initial carbohydrate-mediated binding between stationary and rapidly moving cells has to be only sufficiently tight to allow time for the subsequent action of protein adhesion systems that are defective in cancer cells.

On the other hand, it is clear that cancer metastasis development requires the participation of host cells, e.g., leukocytes and platelets (see above). These considerations lead us to the conclusion that the formation of "multicellular metastatic units" in the blood stream through homotypic and heterotypic carbohydrate-mediated cancer cell adhesion is a prerequisite of cancer metastasis. These units consist of cancer cells, leukocytes and platelets and site specificity of cancer metastasis may be determined by subset(s) of leukocytes involved in the formation of the "multicellular metastatic unit". Those particular subsets of leukocytes may serve as "carrier cells" targeting the metastatic deposit(s) at the specific site(s) of secondary tumor formation. The initial accumulation of cancer cells in regional lymph nodes during metastasis is in good aggreement with this model, since priming by specific antigens may alter the surface phenotype to enable selective recirculation of lymphocyte subset(s) to the particular type of lymph nodes where the specific antigen was first encountered (see above).

There are a number of observations that show that damage of host tissue due to inflammation or trauma may lead to increased attachment and survival of tumor cells at the site of injury, and enhancement of distant metastases at the site of inflammation or trauma was observed.[111,131,132] It has been found that patients often develop metastases at the site of surgical incisions, even those far from the place(s) of tumor resection.[110] Sugarbaker and Ketcham[133] have suggested that the sites of injury contain blood vessel damage that may allow extravasation of a greater number of tumor cells. In addition to the mechanism discussed, inflammatory responses may aid in invasion and metastasis by a number of different factors, such as lysosomal enzyme release from accumulating neutrophils, leukocytes or macrophages.[105] The enhancement of cancer metastases development at a site of injury or inflammation is in agreement with a concept of multicellular metastatic units and the conductor-like role of specific leukocyte subsets in metastasis since it is well known that leukocytes accumulate selectively at the site(s) of injury or inflammation.

References

1. Fidler, I.J., Gersten, D.M., Hart, I.R. (1978) The biology of cancer invasion and metastasis. Adv. Cancer Res. 28:149-250.
2. Liotta, L.A. (1986) Tumor invasion and metastasis—role of the extracellular matrix: Rhoads Memorial Award Lecture. Cancer Res. 46:1-7.
3. Schirrmacher, V. (1985) Cancer metastasis: Experimental approaches, theoretical concepts and impacts for treatment strategies. Adv. Cancer Res. 43:1-73.
4. Dano, K., Andreasen, P. A., Grondahl-Hansen, J., Kristensen, B.P., Nielsen, L.S., Skriver, L. (1985) Plasminogen activatiors, tissue degradation and cancer. Adv. Cancer Res. 44:139-266.
5. Mullins, D.E., Rohrlich, S.T. (1983) The role of proteinases in cellular invasiveness. Biochim. Biophys. Acta. 695:177-214.
6. Reich, E. (1978) Activation of plasminogen: a general mechanism for producing localized extracellular proteolysis. In: Molecular basis of biological degradation processes. Editors: R.D. Berlin, H.Herrmann, I.H. Lepow and J.M. Tanzer, Academic Press, New York, p. 155-169.
7. Reich, R.. Thompson, E., Iwamoto, Y., Martin, G.R., Deason, J.R., Fuller, G.C., Miskin, R. (1987) Effects of inhibitors of plasminogen activator, serine proteinases and collagenase IV on the invasion of basement membrane by metastatic cells in mice and humans. Cancer Res. 48:3307-3312.
8. Mignatti, P., Robbins, E., Rifkin, D.B. (1986) Tumor invasion through the human amniotic membrane: requirement for a proteinase cascade. Cell 47:487-498.
9. Ossowski, L. (1988) Plasminogen activator dependent pathways in the dissemination of human tumor cells in the chick embryo. Cell 52:321-328.
10. Ossowski, L., Reich, E. (1983) Antibodies to plasminogen activator inhibit human tumor metastasis. Cell 35:611-619.
11. Axelrod, J.H, Reich, R., Miskin, R. (1989) Expression of human recombinant plasminogen activators

enhances invasion and experimental metastasis of H-ras-transformed NIH 3T3 cells. Mol. Cell. Biol. 9:2133-2141.

12. Skriver, L., Larsson, L.-I., Keilberg, V., Nielsene, L.S.,resen, P.B., Kristensen, P., Dano, K. (1984) Immunocytochemical localization of urokinase-type plasminogen activator in Lewis lung carcinoma. J. Cell Biol. 99:753-757.

13. Hearing, V.J., Law, L.W., Corti, A., Appella, E., Blasi, F. (1988) Modulation of metastatic potential by cell surface urokinase of murine melanoma cells. Cancer Res. 48:1270-1278.

14. Mueller-Kleiser, W. (1987) Multicellular spheroids. A review on cellular aggregates in cancer research. J. Cancer Res. Clin. Oncol. 113, 101-122.

15. Sutherland, R.M. (1988) Cell and environment interactions in tumor microregions: the multicell spheroid model. Science. 240: 177-184.

16. Nicolson, G.L. (1982) Cancer metastasis. Organ colonization and the cell-surface properties of malignant cells. Biochim. Biophys. Acta 695:113-176.

17. Nicolson, G.L. (1984) Cell surface molecules and tumor metastasis. Regulation of metastatic phenotypic diversity. Exp. Cell Res. 150:3-22.

18. Roos, E. (1983) Cellular adhesion, invasion and metastasis. Biochim. Biophys. Acta 738:263-284.

19. Fidler, I.J. (1973) The relationship of embolic homogeneity, number, size and viability to the incidence of experimental metastasis. Eur. J. Cancer 9:223-227.

20. Glaves, D. (1983) Correlation between circulating cancer cells and incidence of metastases. Br. J. Cancer 48:665-673.

21. Liotta, L.A., Kleinerman, J., Saidel, G.M. (1976) The significance of hematogenous tumor cell clumps in the metastatic process. Cancer Res. 36:889-894.

22. Varani, J., Orr, W., Ward, P.A. (1980) Adhesive characteristics of tumor cell variants of high and low tumorigenic potential. J. Natl. Cancer Inst. 64:1173-1178.

23. Winkelhake, J.L., Nicolson, G.L. (1976) Determination of adhesive properties of variant metastatic melanoma cells to BALB/3T3 cells and their virus-transformed derivatives by a monolayer attachment assay. J. Natl. Cancer Inst. 56:285-291.

24. Fidler, I.J. (1973) Selection of successive tumour lines for metastasis. Nat. New Biol. 242:148-149.

25. Nicolson, G.L., Brunson, K.W., Fidler, I.J. (1978) Specificity of arrest, survival, and growth of selected metastatic variant cell lines. Cancer Res. 38:4105-4111.

26. Fidler, I.J., Bucana, C. (1977) Mechanism of tumor cell resistance to lysis by syngeneic lymphocytes. Cancer Res. 37:3945-3956.

27. Gasic, G.L., Gasic, T.B., Galanti, N., Johnson, T., Murphy, S. (1973) Platelet-tumor-cell interactions in mice. The role of platelets in the spread of malignant disease. Int. J. Cancer 11:704-718.

28. Nicolson, G.L., Winkelhake, J.L. (1975) Organ specificity of blood-borne tumour metastasis determined by cell adhesion? Nature (Lond,), 255:230-232.

29. Raz, A., Bucana, C., McLellan, W., Fidler, I.J. (1980) Distribution of membrane anionic sites on B16 melanoma variants with differing lung colonising potential. Nature (Lond)., 284:363-364.

30. Damsky, C.H., Knudsen, K.A., Buck, C.A. (1983) In: Biology of Glycoproteins. Editor: R. Ivatt. NY: Academic Press. pp. 1-64.

31. Lotan, R., Raz, A. (1983) Low colony formation in vivo and in culture as exhibited by metastatic melanoma cells selected for reduced homotypic aggregation. Cancer Res. 43:2088-2093.

32. Lotan, R., Lotan, D., Raz, A. (1985) Inhibition of tumor cell colony formation in culture by a monoclonal antibody to endogenous lectins. Cancer Res. 45:4349-4353.

33. Lotan, R., Cavvalero, D., Lotan, D., Raz, A. (1989) Biochemical and immunological characterization of K-1735P melanoma galactoside-binding lectins and their modulation by differentiation inducers. Cancer Res. 49:1261-1268.

34. Meromsky, L., Lotan, R., Raz, A. (1986) Implications of endogenous tumor cell surface lectins as mediators of cellular interactions and lung colonization. Cancer Res., 46:5270-5275.

35. Raz, A., Lotan, R. (1981) Lectin-like activities associated with human and murine neoplastic cells. Cancer Res. 41:3642-3647.

36. Barondes, S.H. (1984) Soluble lectins: a new class of extracellular proteins. Science, 23:1259-1264.

37. Briles, E.B., Gregory, W., Fletcher, P., Kornfeld, S. (1979) Vertebrate lectins, Comparison of properties of beta-galactoside-binding lectins from tissues of calf and chicken. J. Cell. Biol. 81:528-537.

38. Cerra, R.F., Haywood-Reid, P.L., Barondes, S.H. (1984) Endogenous mammalian lectin localized extracellularly in lung elastic fibers. J. Cell. Biol. 98:1580-1589.

39. Nowak, T.P., Kobiler, D., Roel, L.E., Barondes, S.H. (1977) Developmentally regulated lectin from embryonic chick pectoral muscle. Purification by affinity chromatography. J. Biol. Chem. 252:6026-6030.

40. Zalik, S.E., Thompson L.W., Ledsham, I.M. (1987) Expression of an endogenous galactose-binding lectin in the early chick embryo. J. Cell. Sci. 88:483-493.

41. Cook, G.M., Zalik, S.E., Milos, N., Scott, V.A. (1979) A lectin which binds specifically to beta-D-galactoside groups is present at the earliest stages of chick embryo development. J. Cell. Sci. 38:293-304.

42. Drickamer, K. (1988) Two distinct classes of carbohydrate-recognition domains in animal lectins. J. Biol. Chem. 263:9557-9560.

43. Harrison, F.L., Chesterton, C.J. (1980) Factors mediating cell—cell recognition and adhesion. Galaptins, a recently discovered class of bridging molecules. FEBS Lett. 122:157-165.

44. Monsigny, M., Kieda, C., Roche, A.C. (1983) Biol. Cell. 47:95-110.

45. Raz, A., Lotan, R. (1987) Endogenous galactoside-binding lectins: A new class of functional tumor cell surface molecules related to metastasis. Cancer Metast.

Rev. 6:433-452.

46. Regan, L.J., Dodd, J., Barondes, S.H., Jessel, T.M. (1986) Selective expression of endogenous lactose-binding lectins and lactoseries glycoconjugates in subsets of rat sensory neurons. Proc. Natl. Acad. Sci. USA, 83:2248-2252.

47. Jia, S., Mee, R., Morford, G., Argwal, N., Voss, P., Moutsatsos, I., Wang, J. (1987) Carbohydrate-binding protein 35: molecular cloning and expression of a recombinant polypeptide with lectin activity in *Escherichia coli*. Gene, 60:197-204.

48. Moutsatsos, I.K., Wade, M., Schindler, M., Wang, J.L. (1987) Endogenous lectins from cultured cells: nuclear localization of carbohydrate-binding protein 35 in proliferating 3T3 fibroblasts. Proc. Natl. Acad.Sci. USA, 84:6452-6456.

49. Hinck, A., Wrenn, D., Mecham, R.P., Barondes, S.H. (1988) The elastin receptor: A galactoside-binding protein. Science 239:1539-1541.

50. Allen, H.J., Karakousis, C., Piver, M.S., Gamarra, M., Nava, H., Forsyth, B., Matecki, B., Jazayeri, A., Sucato, D., Kisailus, E., DiCioccio, R. (1987) Galactoside-binding lectin in human tissues. Tumor Biol. 8:218-229.

51. Carding, S.R., Thorpe, S.J., Thorpe, R., Feizi, T. (1983) Transformation and growth related changes in levels of nuclear and cytoplasmic proteins antigenically related to mammalian beta-galactoside-binding lectin. Biochem. Biophys. Res. Commun. 127:680-686.

52. Gabius, H.-J., Engelhardt, R., Cramer, F. (1986) Endogenous tumor lectins: Overview and perspectives. Anticancer Res. 6:573-578.

53. Gitt, M.A., Barondes, S.H., (1986) Evidence that a human soluble beta-galactoside-binding lectin is encoded by a family of genes. Proc. Natl. Acad. Sci. USA, 83:7603-7607.

54. Raz, A., Meromsky, L., Carmi, P., Karakash, R., Lotan, D., Lotan, R. (1984) Monoclonal antibodies to endogenous galactose-specific tumor cell lectins. EMBO J. 3:2979-2983.

55. Raz, A., Meromsky, L., Lotan, R. (1986) Differential expression of endogenous lectins on the surface of nontumorigenic, tumorigenic, and metastatic cells. Cancer Res., 46:3667-3672.

56. Raz, A., Avivi, A., Pazarini, G., Garmi, P. (1987) Cloning and expression of cDNA for two endogenous UV-2237 fibrosarcoma lectin genes. Exp. Cell Res., 173:109-116.

57. Raz, A., Meromsky, L., Zvibel, I., Lotan, R. (1987) Transformation-related changes in the expression of endogenous cell lectins. Int. J. Cancer, 39:353-360.

58. Raz, A, Carmi. P., Pazarini, G. (1988) Expression of two different endogenous galactoside-binding lectins sharing sequence homology. Cancer Res., 48:645-649.

59. Gabius, H.-J., Engelhardt, R., Cramer, F., Batge, R., Nagel, G.A. (1985) Pattern of endogenous lectins in a human epithelial tumor. Cancer Res. 45:253-257.

60. Irimura, T., Matsushita, Y., Sutton, R.C., Carralero, D., Ohannesian, D.W., Cleary, K.R., Ota, D.M., Nicolson, G.L., Lotan, R. (1991) Increased content of an endogenous lactose-binding lectin in human colorectal carcinoma progressed to metastatic stages. Cancer Res. 51:387-393.

61. Gasic, G.J., Gasic, T.B., Stewart, C.C. (1968) Antimetastatic effects associated with platelet reduction. Proc. Natl. Acad. Sci USA 61:46-52.

62. Cavanaugh, P.G., Sloane, B.F., Honn, K.V. (1988) Role of the coagulation system in tumor cell-induced platelet aggregation and metastasis. Haemostasis 18:37-46.

63. Grignani, G., Jamieson, G.A. (1988) Platelets in tumor metastasis: Generation of adenosine diphosphate by tumor cells is specific but unrelated to metastatic potential. Blood, 71:844-849.

64. Tsuruo, T., Watanabe, M., Oh-hara, T. (1989) Stimulation of the growth of metastatic clones of mouse colon adenocarcinoma 26 in vitro by platelet-derived annagi ion.

65. Ugen, K.E., Mahalingam, M., Klein, P.A., Kao, K.-J. (1988) Inhibition of tumor cell-induced platelet aggregation and experimental tumor metastasis by the synthetic Gly-Arg-Gly-Asp-Ser peptide. J. Natl. Cancer Inst. 80:1461-1466.

66. Glinsky, G.V. (1990) Immunoselective hypothesis of tumor progression. Role aberrant glycosylation, anti-carbohydrate antibodies, extracellular glycomacromolecules and glycoamines. J. Tumor Marker Oncology. 5:206.

67. Glinsky, G.V. (1990) Glycoamines, aberrant glycosylation and cancer: a new approach to the understanding of molecular mechanism of malignancy., In: Molecular Oncology. Oncodevelopment proteins and clinical applications. XVIIIth meeting of the International Society for Oncodevelopmental Biology and Medicine. Abstract Book, Moscow, USSR, September 23-27, 1990, p.7.

68. Glinsky, G.V., Semyonova-Kobzar, R.A., Berezhnaya, N.M. (1990) Modification of cellular adhesion, metastasizing and immune response by glycoamines: implication in the pathogenetical role and potential therapeutic application in tumoral disease. J. Tumor Marker Oncology. 5:231.

69. Glinsky, G.V. (1992) Glycoamines: Structural-Functional characterization of a new class of human tumor markers. In: Serological Cancer Markers. Editor: S. Sell. The Humana Press., Totowa, NJ, Chapter 11, p. 233-260.

70. Glinsky, G.V. (1992) The blood group antigens (BGA)-related glycoepitopes. A key structural determinants in immunogenesis and cancer pathogenesis. Critical Reviews in Oncology/Hematology 12:151-166.

71. Evans, C.W. (1991) The metastatic cell: Behavior and biochemistry. Chapman and Hall., NY, 551 pp.

72. Steeg, P.S. (1991) Tumor metastasis: before the revolution. Cell 66:835-836.

73. Hakomori, S.I. (1988) In: Altered glycosilation in tumor cells. Editors: Ch. L. Reading, S.I. Hakomori, D.M. Marcus. New York: Alan R. Liss, p. 207-212.

74. Hakomori, S.I. (1989) Aberrant glycosylation in

tumors and tumor-associated carbohydrate antigens. Adv. Cancer Res. 52, 257-331.

75. Hakomori, S.I. (1985) Aberrant glycosylation in cancer cell membranes as focused on glycolipids: overview and perspectives. Cancer Res. 45, 2405-2414.

76. Fenderson, B.A., Andrews, P.W., Nudelman, E., Clausen, H., Hakomori, S.I. (1987) Glycolipid core structure switching from globo to lacto-and ganglio-series during retinoic acid-induced differentiation of TERA-2-derived human embryonal carcinoma cells. Dev. Biol. 122, 21-34.

77. Springer, G.F. (1984) T and Tn, general carcinoma autoantigens. Science, 224 1198-1206.

78. Springer, G.F., Cheinsong-Popov, R., Schirrmacher, V., Desai, P.R., Tegtmeyer, H. (1983) Proposed molecular basis of murine tumor cell-hepatocyte interaction. J. Biol. Chem. 258, 5702-5706.

79. Fenderson, B.A., Eddy, E.M., Hakomori, S.I. (1990) Glycoconjugate expression during embryogenesis and its biological significance. BioEssay 12, 173-79.

80. Lindenberg, S., Sundberg, K., Kimber, S.J., Lundblad, A. (1988) The milk oligosaccharide, lacto-N-fucopentaose I, inhibits attachment of mouse blastocysts on endometrial monolayers. J. Reprod. Fert. 83, 149-158.

81. Abe, K., Hakomori, S.I., Ohshiba, S. (1986) Differential expression of difucosyl type 2 chain (LeY) defined by monoclonal antibody AH6 in different locations of colonic epithelia, various histological types of colonic polyps, and adenocarcinomas. Cancer Res. 46, 2639-2644.

82. Itzkowitz, S.H., Shi, Z.R., Kim, Y.S. (1986) Heterogeneous expression of two oncodevelopmental antigens, CEA and SSEA-1, in colorectal cancer. Histochem. J. 18:155-163.

83. Itzkowitz, S.H., Yuan, M., Gukushi, Y., Palekar, A., Phelps, P.C., Shamsuddin, A.M., Trump, B.F., Hakomori, S.I., Kim, Y.S. (1986) Lewisx- and sialylated Lewisx-related antigen expression in human malignant and nonmalignant colonic tissues. Cancer Res. 46, 2627-2632.

84. Itzkowitz, S.H., Yuan, M., Ferrell, L.D., Palekar, A., Kim, Y.S. (1986) Cancer-associated alterations of blood group antigen expression in human colorectal polyps. Cancer Res. 46, 5976-5984.

85. Itzkowitz, S.H., Yuan, M., Montgomery, C.K., Kjeldsen, T., Takahashi, H.K., Bigbee, W.L., Kim, Y.S. (1989) Expression of Tn, sialosyl-Tn, and T antigens in human colon cancer. Cancer Res. 49, 197-204.

86. Kim, Y.S., Yuan, M., Itzkowitz, S.H., Sun, Q., Kaizu, T., Palekar, A., Trump, B.F., Hakomori, S.I. (1986) Expression of LeY and extended LeY blood group-related antigens in human malignant, premalignant and nonmalignant colonic tissues. Cancer Res. 46, 5985-5992.

87. Hakomori, S.-I. (1991) Possible functions of tumor-associated carbohydrate antigens. Current Opinion in Immunology. 3:646-653.

88. Hakomori, S.-I. (1992) Possible new directions in cancer therapy based on aberrant expression of glycosphingolipids: Antiadhesion and ortho-signaling therapy. Cancer Cells (in press).

89. Itzkowitz, S.H., Bloom E.J., Kokal, W.A., Modin, G., Hakomori, S., Kim, Y.S. (1990) Sialosyl-Tn: A novel mucin antigen associated with prognosis in colorectal cancer patients. Cancer 66:1960-1966.

90. Hoff, S.D., Irimura, T., Matsushita, Y., Ota, D.M., Cleary, K.R., Hakomori, S. (1990) Metastatic potential of colon carcinoma: Expression of ABO/Lewis-related antigens. Arch. Surg. 125:206-209.

91. Castronovo, V., Colin, C., Parent, B., Foidart, J.M., Lambotte, R., Mahieu, P. (1989) Possible role of human natural anti-Gal antibodies in the natural antitumor defense system. J. Natl. Cancer Inst. 81:212-216.

92. Hausman, R.E. (1983) Increase in homotypic aggregation of metastatic Morris hepatoma cells after fusion with membranes from non-metastatic cells. Int. J. Cancer 32:603-608.

93. Viadana, E., Au, K.L. (1975) Patterns of metastases in adenocarcinomas of man. An autopsy study of 4,728 cases. J. Med. 6:1-14.

94. Viadana, E., Bross, I.D.J., Pickren, I.W. (1973) An autopsy study of some routes of dissemination of cancer of the breast. Br. J. Cancer 27:336-340.

95. Bross, I.D.J., Viadana, E., Pickren, I.W. (1975a) Do generalized metastases occur directly from the primary? J. Chron. Dis. 28:149-159.

96. Bross, I.D.J., Viadana, E., Pickren, I.W. (1975b) The metastatic spread of myeloma and leukemias in men. Virchows Arch. A. Path. Anat. Hist. 365:91-101.

97. Onuigbo, W.I.B. (1974) Oncology, 30:294-303.

98. Weiss, L. (1967) In: The Cell Periphery, Metastasis and other Phenomena. North Holland, Amsterdam, pp. 339-341.

99. Ewing, J. (1928) Neoplastic Diseases, 3rd edition. W.B. Sanders, Philadelphia, PA

100. Paget, S. (1889) Lancet, i:571-573.

101. Weiss, L. (1977) A pathobiologic overview of metastasis. Semin. Oncol. 4:5-19.

102. Sugarbaker, E.V. (1979) Cancer metastasis: a product of tumor-host interactions. Curr. Probl. Cancer 3:3-59.

103. Fidler, I.J., Nicolson, G.L. (1981) Cancer Biol. Rev. 2:171-234.

104. Hart, I.R., Fidler, I.J. (1981) The implications of tumor heterogeneity for studies on the biology of cancer metastasis. Biochim. Biophys. Acta. 651:37-50.

105. Sugarbaker, E.V. (1981) Cancer Biol. Rev. 2:235-278.

106. Elliot, R.H.E., Jr., Frantz, V.R. (1960) Ann. Surg. 151:551-561.

107. Prout, G.R., Jr. (1973) In: Cancer Medicine., Editors: Holland, J.F., Frei, E. Lea and Febiger, Philadelphia, PA, pp. 1680-1694.

108. Hansen, H.H., Muggia, F.M. (1972) Staging of inoperable patients with bronchogenic carcinoma with special reference to bone marrow examination and peritoneoscopy. Cancer 30:1395-1401.

109. Patel, J.K., Didolkar, M.S., Pickren, J.W., Moore, R.H. (1978) Metastatic pattern of malignant melanoma. A study of 216 autopsy cases. Am. J. Surg. 135:807-810.

110. Der Hagopian, R.P., Sugarbaker, E.V., Ketcham, A. (1978) Inflammatory oncotaxis. J. Am. Med. Assoc. 240:374-375.

111. Sugarbaker, E.V., Cohen, A.M., Ketcham, A.S. (1971) Do metastases metastasize? Ann. Surg. 174:161-166.

112. Hart, I.R., Fidler, I.J. (1980) Role of organ selectivity in the determination of metastatic patterns of B16 melanoma. Cancer Res. 40:2282-2287.

113. Kinsey, D.L. (1960) Cancer 13:674-676.

114. Griffiths, J.D., Salsburg, A.J. (1963) Br. J. Cancer 17:546-557.

115. Roos, E., Dingemans, K.P. (1979) Mechanisms of metastasis. Biochim. Biophys. Acta. 560:135-166.

116. Poste, G., Fidler, I.J. (1980) The pathogenesis of cancer metastasis. Nature, 283:139-146.

117. Fidler, I.J. (1975) Biological behavior of malignant melanoma cells correlated to their survival in vivo. Cancer Res., 35:218-224.

118. Fidler, I.J. (1978) Tumor heterogeneity and the biology of cancer invasion and metastasis. Cancer Res., 38:2651-2660.

119. Fidler, I.J., Nicolson, G.L. (1976) Organ selectivity for implantation survival and growth of B16 melanoma variant tumor lines. J. Natl. Cancer Inst. 57:1199-1201.

120. Fidler, I.J., Nicolson, G.L. (1977) Fate of recirculating B16 melanoma metastatic variant cells in parabiotic syngeneic recipients. J. Natl. Cancer Inst. 58:1867-1872.

121. Nicolson, G.L. (1978) BioScience 28:441-447.

122. Nicolson, G.L., Brunson, K.W. (1977) Gann. Monogr. Cancer Res. 20:15-24.

123. Del Regato, J.A. (1977) Pathways of metastatic spread of malignant tumors. Semin. Oncol. 4:33-38.

124. Proctor, J.W. (1976) Rat sarcoma model supports both soil seed and mechanical theories of metastatic spread. Br. J. Cancer 34:651-654.

125. Einhorn, L.H., Burgess, M.A., Vallejos, C., Body, G.P., Gutterman, J., Mavligit, G., Hersh, E., Luce, J.K., Frei, E., Freireich, E.J., Gottlieb, J.A. (1974) Prognostic correlations and response to treatment in advanced metastatic malignant melanoma. Cancer Res. 34:1995-2004.

126. Stehlin, J.S., Hills, W.J., Rufino, C. (1967) Disseminated melanoma. Biologic behavior and treatment. Arch. Surg. 94:495-501.

127. Fisher, B., Fisher, E.R. (1967) The organ distribution of disseminated 51 Cr-labeled tumor cells. Cancer Res. 27:412-420.

128. Fisher, B., Fisher, E.R. (1966) The interrelationship of hematogenous and lymphatic tumor cell dissemination. Surg. Gynecol. Obstet. 122:791-798.

129. Springer, T.A., Lasky, L.A. (1991) Sticky sugars for selectins. Nature 349:196-197.

130. Miyake, M., Hakomori, S. (1991) A specific cell surface glycoconjugate controlling cell motility: Evidence by functional monoclonal antibodies that inhibit cell motility and tumor cell metastasis. Biochemistry 30:3328-3334.

131. Smith, R.R., Thomas, L.B., Hilbers, A.W. (1958) Cancer 11:53-62.

132. Fisher, B., Fisher, E.R., Feduska, N. (1967) Trauma and the localization of tumor cells. Cancer 20:23-30.

133. Sugarbaker, E.V., Ketcham, A.S. (1977) Mechanisms and prevention of cancer dissemination: An overview. Semin. Oncol. 4:19-32.

134. Zeidman, I. (1961) Cancer Res. 21:38-39.

135. Fisher, B., Fisher, E.R. (1966) Transmigration of lymph nodes by tumor cells. Science 152:1397-1398.

136. Kurokawa, Y. (1970) Experiments on lymph node metastasis by intralymphatic inoculation of rat ascites tumor cells, with special reference to lodgement, passage and growth of tumor cells in lymph nodes. Gann. 61:461-470.

CHAPTER 7

GLYCODETERMINANTS OF MELANOMA CELL ADHESION AS MODEL COMPOUNDS OF ANTIMETASTATIC DRUG DESIGN

Adhesion properties of cancer cells have been recognized as very significant biological features of malignant tumor related to metastasis, and carbohydrates and carbohydrate-binding proteins have been identified as key structural determinants involved in tumor cell adhesion. The antimetastatic activity of several natural and synthetic cell adhesion inhibitors has been shown on an experimental model of metastasis. Recently, the possibility of using the tumor cell aggregation glycodeterminants themselves as a tool for antimetastatic therapy of human cancer has been suggested.[1,2] This chapter will be focused on the analysis of current data on the biological characterization of glycodeterminants of melanoma cell adhesion and the prospect of antimetastatic drug development for malignant melanoma. The role in these processes of an extracellular serum biomacromolecule carrying and/or specifically recognizing the glycodeterminants of cancer cell aggregation will be discussed. A model for design and application for treatment of human cancer of natural and synthetic cell adhesion inhibitors with biospecificity for glycodeterminants of tumor cell aggregation will be considered.

INTRODUCTORY REMARKS

Metastasis is one of the most important malignant features of human cancer and represents the morphological stage of the generalization of the disease through the body of the tumor host. It is obvious that development of new methods of early detection and prognosis of metastasis, identification of the host predisposition factors for metastasis, as well as discovery of key molecular events in metastatic spread, could result in increasingly successful treatment of human cancer.

Homotypic and heterotypic aggregation properties of tumor cells represent important biological features of malignancy because these properties of transformed cells could determine the metastatic properties of neoplastic tumors.[3-19]

The key structural determinants of the tumor cells that participate in cell recognition, association and aggregation are as carbohydrates and carbohydrate-binding proteins.[3-5,14-18,20-26]

The incidence of malignant melanoma is increasing at a faster rate than any other malignancy, with an estimated 32,000 new cases and 6500 deaths in the United States in 1991.[27] At the current rate, it is estimated that one out of every 105 Americans born in 1991 will develop malignant melanoma compared to an estimated one out of 1500 in 1935.[28] Despite these statistics, the survival rate for malignant melanoma has increased from 40% five-year survival in the 1940s to 80% five-year survival in the 1980s.[29,30] Since adjunctive therapies beyond surgical excision have not improved over this period, the improved survival is attributed to earlier diagnosis and excision. It has long been noted by both Clark[31] and Breslow[32] that there was an inverse relationship between depth of melanoma invasion and survival. Patients with melanomas less than 0.76 mm thick have nearly a 100% five-year survival compared to about 40% five-year survival for melanomas thicker than 4.00 mm.[33] Hence, the current improvement in survival is probably due to enhanced patient awareness and removal of melanomas at a thinner stage, and development of new treatment strategy for recurrence and metastatic melanoma.

Malignant melanoma, a tumor of neural crest origin, synthesizes large quantities of gangliosides (G_{M3}, G_{M2}, G_{D3} and G_{D2}) that are expressed on the surface of melanoma cells. It is important that G_{M2} and G_{D3} expression of human melanoma is related to cellular adhesion (see below) and increased G_{M2} and G_{D2} in melanoma is correlated with greater tumorigenicity in nude mice (see below). In vitro tests indicate that free gangliosides are potent inhibitors of immune responses and in vivo studies show an influence of free gangliosides on growth of primary and metastatic tumors (see below). Thus, the carbohydrate units of gangliosides are primary bioaddressing structures in synthetic compounds specifically designed as potential antimetastatic and/or antitumoral agents for human melanoma.

Tumor cell aggregation glycodeterminants, on the other hand, have unique potential as tools for pathogenetic anticancer and antimetastatic therapy with regard to tumor bioselectivity and sensitivity. For example, the disialoganglioside G_{D2} is expressed on the surface of cells of most human melanomas but not on the surface of melanocytes,[34] the ganglioside 9-0-acetyl-G_{D3} is expressed by cells of human melanoma but not by other cells including melanocytes.[35,36] During recent years dramatic progress has been made in our understanding of molecular mechanisms of cell adhesion, particularly in identification and structural-functional characterization of several cellular receptor systems and their ligands participating in a multistep "cascade" cell adhesion reaction.

Glycosphingolipids as Recognition Sites of Cell-Cell Interactions

Cell-specific glycosylation patterns and their alterations during development, cell differentiation and oncogenesis are well-documented.[52] Carbohydrate-carbohydrate interactions could provide the initial step of cell type-specific recognition prior to cell type-nonspecific adhesion.[37,38] One of the possible functions of glycosphingolipids (GSL) at the cell surface is as a recognition site for determining specificity of cell-cell interactions since 1) dramatic changes in GSL composition are associated with cell differentiation, development, and oncogenesis;[52] 2) specific types of GSL or their oligosaccharide sequences inhibit cell recognition occurring during morphogenesis and histogenesis;[53-57] and 3) recently the evidence has been presented for specific interaction between two GSL, sialosyllactosyl-ceramide (G_{M3}) and gangliotriaosylce-ramide (G_{G3}) and clear heterotypic specific cellular recognition and interaction between melanoma cells and lymphoma cells expressing G_{M3} and G_{G3}, respectively.[38]

Experimental evidence supports the concept that some of the BGA-related glycodeterminants which have been identified earlier as tumor associated carbohydrate antigens (TACA) function as key adhesion molecules.[1,2,48] The recent studies have shown that cell adhesion through carbohydrate-carbohy-drate or carbohydrate-selectin interactions occur at an early initial stage of the "cascade" multistep cell adhesion mechanism, and this reaction is prerequisite for integrin- or immunoglobulin-based adhesion.[49-51] Usually cells coexpress on their surface the multiple components involved in "cascade" cell adhesion , and thus this multistep adhesion reaction could be triggered by initial carbohydrate-carbohydrate or carbohydrate-selectin interaction. Those conclusions have been made on the basis of study of B16 melanoma cell adhesion[2,49,50] and investigation of leukocytes' role on a selectin.[51] Evidence has been presented that specific glycosplingolipid-glycosphingolipid interaction initiates cell-cell adhesion and may cooperate synergistically with other cell-adhesion systems such as those involving integrins.[2,48-50]

Ganglioside Expression on the Surface of Melanoma Cells

Gangliosides are membrane-bound sialic acid-containing glycosphingolipids that are found in high concentration on neural tissues, and they are greatly increased in neuroectodermal tumors.[58-63] Malignant melanoma, a tumor of neural crest origin, synthesizes large quantities of gangliosides (G_{M3}, G_{M2}, G_{D3}, G_{D2} and alkali-labile gangliosides). The disialoganglioside G_{D2} is expressed on the surface of cells of most human melanoma but not on the surface of melanocytes[34] and ganglioside 9-0-acetyl-G_{D3} is expressed on the surface of cells of human melanoma but not by other cells including melanocytes.[35,36] The relationship has been investigated between various clinical factors (age, sex, site, stage, tumor size, pigmentation, histopathologic type of primary tumor, chemosensitivity and prognosis).[64] It was found that G_{M3} positively correlated with a good prognosis in both biopsy and cultured melanoma. When radiosensitivity or chemosensitivity of human melanoma cells was tested in vitro using the human tumor colony forming assay those cells expressing a low content of G_{M3} or a high content of G_{D2} correlated proportionally with the sensitivity to radiation and phase-specific

drug treatment.[65] The total ganglioside amount (ranging from 33 to 302 mg/g wet wt of tissue) as well as the distribution (pattern) of each ganglioside were widely heterogenous in both biopsied and cultured melanoma.[66] It has been reported on the other hand that neurofibroma, a benign tumor, has the same ganglioside profiles among different patients tumors, whereas the malignant tumors, neurofibrosarcomas, exhibited different ganglioside patterns among different patients tumors.[67] Heterogeneous expression of melanoma antigens has been demonstrated by using immunologic methods with monoclonal antibodies.[68-72]

In contrast, comparison of different sites of metastasis revealed only a slight difference in the amount of G_{D3} between tumor biopsy specimens from lymph nodes and subcutaneous tissues. Despite wide variations in the ganglioside composition of human melanoma in different individuals, when several tumor specimens were obtained from the same individual they expressed similar ganglioside profiles in spite of differences in the site of tumor and date of biopsy.[64] When patients had more than two tumors in different tissues the ganglioside profiles changed only slightly at different times in the same patient. Two samples were derived from the same patient at a two and a half year interval as a stage II middle-sized amelanotic tumor from lymph node and as a stage III large-sized melanotic tumor from the mediastinum, but ganglioside profiles were basically alike.[64] Using cultured melanoma, three cell lines from different tumors in the same patient have been investigated. The ganglioside profiles were nearly identical in the three cell lines: one was obtained from a brain metastasis in August 1973, the second was obtained from a subcutaneous metastasis 14 months later, and the third was obtained from a sigmoid colon metastasis 15 months later.[64] These results are very important since they indicate that some of phenotypic properties of melanoma cells associated with aberrant glycosylation are stable and preserved regardless of tumor progression. A similar conclusion has been made regarding human carcinoma.[1] Thus, these phenotypic properties of tumor cells that are associated with aberrant glycosylation and involving tumor cell aggregation glycodeterminants should be recognized as important from the point of view of their role in cancer pathogenesis and as a target for anticancer drug development.

The Disialoganglioside G_{D3} is the major sialoglycosplingolipid of human melanoma cells.[60,73] Monoclonal antibodies (mAbs) against G_{D3} are currently in use in clinical trials for the serotherapy of human melanoma.[74-76] Among the gangliosides expressed as cell surface antigens in human melanoma, 9-0-acetyl-G_{D3} is of particular interest because of its reported restriction to melanoma[35,36,77] and consequently it is a candidate antigen for vaccine development.[78]

G_{M3} and G_{D3} are major gangliosides in many human adult tissues, conversely, G_{M2} is expressed widely in many malignant tumors,[67,79,80] and G_{D2} is expressed in malignant cells of neural crest origin.[67,80,81] Thus, in terms of specificity, application of mAbs may be optional with the use of antibodies against G_{M2} and G_{D2}. An obvious problem for the use of antibodies in treatment against G_{M3} is the fact that this ganglioside is widely distributed as a major ganglioside on various normal human tissues, including retina, kidney, and adrenal medulla.[82,83] However, successful immunotherapy by mouse IgG anti-G_{D3} antibodies has been reported.[74] None of the patients who received antibodies (up to 523 mg) developed neurological or renal toxicities, and objective clinical responses were demonstrated in 3 of 11 stage III melanoma patients. Crypticity of antigenic determinants of G_{D3} on cell membranes of normal cells is one of the possible explanation of observed unresponsiveness of normal tissues to anti-G_{D3} antibody treatment.

Shedding of Gangliosides by Melanoma Cells: The Pathophysiological Significance

Tumor cells, particularly melanoma cells, shed gangliosides.[36, 84-89] These findings suggest that gangliosides could be useful humoral tumor markers. A four-fold increase in G_{D3} level was observed in sera of the melanoma patients.[90] Schulz et al,[81] Ladisch et al[92,93] and Yamanaka et al[94] have observed

elevated amounts (up to 40 times more) of circulating ganglioside G_{D2} in serum of neuroblastoma patients. Efforts to monitor antibodies to G_{D2} and 9-0-acetyl GD3 in the sera of melanoma patients have been reported.[95,96] Serum levels of G_{D2} and G_{D3} monitored by sensitive thin-layer chromatography (TLC) and immunostaining were increased approximately six-fold and five-fold, respectively, in patients with disseminated melanoma, compared with healthy adults.[97] Finally, a correlation has been observed between decreased progression and serum levels of a melanoma-associated proteoglycan antigen,[98] a colon carcinoma-associated ganglioside,[99] an ovarian carcinoma antigen[100] and of G_{M2} in hepatoma patients.[101]

A number of in vitro studies have been performed to examine whether gangliosides, which are highly expressed on the cell surface of melanoma cells, are being shed into the culture medium. Measurable quantities of gangliosides G_{M3} and in particular G_{D3} were shed by melanoma cells.[89] Shedding of other gangliosides has been demonstrated by various investigators. The shedding of G_{M2} and G_{D2},[85] G_{M3}[86] and O-acetyl G_{D3}[36] gangliosides from human melanoma cell lines into culture medium has been published. Nilsson et al[87] reported enhancement of fucosyl-G_{M1} in the culture medium of small-cell lung carcinoma cell lines. MAb FHG was used to detect a novel ganglioside in the serum of patients with various tumors.[102] In the serum of some patients the antigen levels were significantly increased, but some patients with high antigen expression in breast cancer tissue had no elevated levels in their serum. Thus, the serum antigen levels do not reflect the degree of antigen expression in tumor tissues, and a number of additional factors affecting the circulatory level of glycolipid and other tumor-associated antigens must be recognized. The amount of gangliosides shed in the local extracellular environment of the tumor could yield ganglioside concentrations which have been shown to be immunosuppresive in vitro.[84] Gangliosides inhibit in vitro function of natural killer cells, monocytes and lymphocytes[84,103-105] some gangliosides suppress the function of T-helper cells [106] and inhibit proliferative responses of interleukin-2-dependent T-lymphocytes[107] and PHA lymphoblasts.[108] A possible role of gangliosides in specific cellular interactions, differentiation, communication and division has been reported.[109-111] There is evidence that certain gangliosides are involved in embryogenesis. MAb R-24 (directed to ganglioside G_{D3}) blocks embryonic kidney development in contrast to many antibodies and probes for cell adhesion molecules, which do not.[112] Depending on the kind of gangliosides, cell types and culture conditions, they can promote or inhibit cell division.[113-116] Several findings suggest involvement of gangliosides in the pathogenesis of malignancy. G_{D3} expression is augmented during malignant transformation of melanocytes.[117-119] Ganglioside mixtures extracted from various mouse tumor cell lines improve growth and mobilization of capillary endothelium and neoplastic cells, and gangliosides from highly tumorigenic lymphoma cells markedly increased tumorigenicity of poorly tumorigenic cells in mice.[120] Free gangliosides enhanced tumor growth and frequency of metastasis in vivo.[121]

GANGLIOSIDES AS STRUCTURAL SITES OF ADHESION OF MELANOMA CELLS

Gangliosides have been found in adhesion plaques and in substrate attachment material.[122,123] G_{D2} and G_{D3} expression of human melanoma was related to cellular adhesion[35] and attachment of melanoma cells to the extracellular matrix proteins (fibronectin) was described to be mediated by G_{D3}.[124] The presence of a high concentration of G_{D2} and G_{D3} in the melanoma cell matrix, which is the main site of cell adhesion, infiltration and invasion, indicates their possible role in the malignant process.[122] Finally, it has been shown that increased amounts of G_{M2} and G_{D2} in melanoma are correlated with greater tumorigenicity in nude mice.[125] Thus, in accordance with our general concept of development of new antimetastatic drugs, the carbohydrate sequences of gangliosides have to be recognized as a primary candidate for structural design of synthetic compounds for treatment of human melanoma. The ganglio-

sides are expressed in large quantities on the cell surface of the melanoma cell. For some of them the expression on human melanoma cells has been shown very specific, and they are involved in melanoma cell adhesion and infiltration as well as in melanoma cell aggregation.

Our studies have shown that glycoamines and other serum macromolecules may play an important role in changing the aggregation properties in vitro and metastasis in vivo of tumor cells, and thus may be involved in the modification of cell recognition, association and aggregation.[1, 39-42] The glycoamines and their synthetic structural analogs (glycoesters and fructosyl amino acids) have revealed strong cell aggregation inhibitory activity in vitro and antimetastatic action in vivo in an experimental model of metastasis. Monoclonal antibodies against cellular lectins that participate in tumor cell aggregation and metastasis, as well as synthetic peptide cell adhesion epitopes, have in vivo antimetastatic activities. Therefore, tumor cell aggregation has strong potential as a biological target for antimetastatic drug development.[1]

CELL SURFACE LECTINS AND SERUM BIOMACROMOLECULES AS A MODIFIER OF CELL-CELL ADHESION

Tumor cell surface lectins play a major role in cellular interactions in vivo that are important for the formation of emboli and for the attachment of circulating tumor cells.[14-18] Tumor cells and tissues contain lectins that are similar to those isolated form normal cells in their sugar-binding specificities, molecular weights, and sequences.[18,126-134] It was found by using monoclonal and polyclonal antilectin antibodies that the lectins were expressed on the cell surface and that their level correlated with malignant transformation and metastatic propensity.[130,131,133] A key role for the tumor-surface lectins in anchorage-independent growth in vitro and in tumor embolization (homotypic and heterotypic aggregation) during the metastatic process in vivo has been suggested.[14-18] The homotypic aggregation of B16 melanoma cells is dependent on the presence of fetal bovine serum.[10,13] One possible explanation for this serum requirement could be that a serum glycomacromolecule(s) mediates intercellular adhesion similar to the action of cell-cell adhesion molecules.

Our studies have shown that the serum glycoamines and other macromolecules may play important roles in changing adhesion and aggregation properties in vitro and metastasis in vivo of cancer cells and thus may be involved in the modification of cell recognition, association and aggregation.[1 39-42] Since carbohydrates and carbohydrate-recognition structures are involved in cell aggregation, we have initiated investigation of structural-functional interrelations of glycoamines, their synthetic structural analogs, and high molecular weight carbohydrate-associated serum tumor markers as a modifiers (inhibitor or stimulator) in vitro of tumor cell aggregation and metastasis in vivo.[1,39-42,136-139]

Thus, the homotypic and heterotypic aggregation properties of tumor cells represent important biological features of malignancy because these properties of transformed cells could determine the metastatic ability of neoplastic tumors. The key structural determinants of the tumor cells that participate in cell recognition, association, and aggregation have been determined as carbohydrates and carbohydrate-binding proteins. Specifically, the BGA (blood group antigens)-related glycoepitopes in human carcinoma and gangliosides in melanoma are directly involved in the homotypic (tumor cells, embryonal cells) and heterotypic (tumor cells-normal cells) formation of cellular aggregates which was demonstrated in different experimental systems. The carbohydrate units of gangliosides should be considered as the key bioaddressing structures of antimetastatic drugs for malignant melanoma.[135] It has been shown that specific carbohydrate units of glycosphingolipids that participate in melanoma cell adhesion, as well as glycosphingolipid-containing liposomes significantly suppressed B16 melanoma lung metastasis.[140] Tumor cell-aggregation properties are particularly important for formation of tumor cell aggregates in the blood stream and their subsequent arrest in the blood vessels of target organs, since the lodgment, attachment, and growth

of blood-borne neoplastic cells depend largely on embolization. The tumor cell aggregation properties and consequently the glycodeterminants of cancer cell aggregation may be also involved in specific cell-cell adhesions required for invasion and extravasation of cancer cells during metastasis.

REFERENCES

1. Glinsky, G.V. (1992) The blood group antigens (BGA) related glycoepitopes. A key structural determinants in immunogenesis and cancer pathogenesis. Critical Reviews in Oncology/Hematology 12:151-166.
2. Hakomori, S.-I. (1992) Possible new directions in cancer therapy based on aberrant expression of glycosphingolipids: Antiadhesion and ortho-signalling therapy. Cancer Cells (in press)
3. Nicolson, G.L. (1982) Cancer metastasis. Organ colonization and the cell-surface properties of malignant cells. Biochim. Biophys. Acta 695:113-176.
4. Nicolson, G.L. (1984) Cell surface molecules and tumor metastasis. Regulation of metastatic phenotypic diversity. Exp. Cell Res. 150:3-22.
5. Roos, E. (1984) Cellular adhesion, invasion and metastasis. Biochim. Biophys. Acta 738:263-284.
6. Fidler, I.J. (1973) The relationship of embolic homogeneity, number, size and viability to the incidence of experimental metastasis. Eur. J. Cancer 9:223-227.
7. Glaves, D. (1983) Correlation between circulating cancer cells and incidence of metastases. Br. J. Cancer 48:665-673.
8. Liotta, L.A., Kleinerman, J., Saidel, G.M. (1976) The significance of hematogenous tumor cell clumps in the metastatic process. Cancer Res. 34:997-1004.
9. Varani, J., Orr, W., Ward, P.A. (1980) Adhesive characteristics of tumor cell variants of high and low tumorigenic potential. J. Natl. Cancer Inst. 64:1173-1178.
10. Winkelhake, J.L., Nicolson, G.L. (1976) Determination of adhesive properties of variant metastatic melanoma cells to BALB/3T3 cells and their virus-transformed derivatives by a monolayer attachment assay. J. Natl. Cancer Inst. 56:285-291.
11. Fidler, I.J., Bucana, C. (1977) Mechanism of tumor cell resistance to lysis by syngeneic lymphocytes. Cancer Res. 37:3945-3956.
12. Gasic, G.L., Gasic, T.B., Galanti, N., Johnson, T., Murphy, S. (1973) Platelet-tumor-cell interactions in mice. The role of platelets in the spread of malignant disease. Int. J. Cancer 11:704-718.
13. Raz, A., Bucana, C., McLellan, W., Fidler, I.J. (1980) Distribution of membrane anionic sites on B16 melanoma variants with differing lung colonising potential. Nature (Lond), 284:363-364.
14. Lotan, R., Raz, A. (1983) Low colony formation in

vivo and in culture as exhibited by metastatic melanoma cells selected for reduced homotypic aggregation.. Cancer Res. 43:2088-2093.

15. Lotan, R., Lotan, D., Raz, A. (1985) Inhibition of tumor cell colony formation in culture by a monoclonal antibody to endogenous lectins. Cancer Res. 45:4349-4353.

16. Lotan, R., Cavvalero, D., Lotan, D., Raz, A. (1989) Biochemical and immunological characterization of K-1735P melanoma galactoside-binding lectins and their modulation by differentiation inducers. Cancer Res. 49:1261-1268.

17. Meromsky, L., Lotan, R., Raz, A. (1986) Implications of endogenous tumor cell surface lectins as mediators of cellular interactions and lung colonization. Cancer Res. 46:5270-5275.

18. Raz, A., Lotan, R. (1981) Lectin-like activities associated with human and murine neoplastic cells. Cancer Res. 41:3642-3647.

19. Raz, A., Lotan, R. (1987) Endogenous galactoside-binding lectins: a new class of functional tumor cell surface molecules related to metastasis. Cancer Metastasis Rev. 6:433-452.

20. Damsky, C.H., Knudsen, K.A., Buck, C.A. (1983) In: Biology of Glycoproteins. Editor: R. Ivatt. NY: Academic Press. pp. 1-64.

21. Barondes, S.H. (1984) Soluble lectins: a new class of extracellular proteins. Science (Wash. D.C.), 23:1259-1264.

22. Fenderson, B.A., Andrews, P.W., Nudelman, E., Clausen, H., Hakomori, S.I. (1987) Glycolipid core structure switching from globo to lacto-and ganglioseries during retinoic acid-induced differentiation of TERA-2-derived human embryonal carcinoma cells. Dev. Biol. 122, 21-34.

23. Fenderson, B.A., Eddy, E.M., Hakomori, S.I. (1990) Glycoconjugate expression during embryogenesis and its biological significance. BioEssay 12:173-179.

24. Lindenberg, S., Sundberg, K., Kimber, S.J., Lundblad, A. (1988) The milk oligosaccharide, lacto-N-fucopentaose I, inhibits attachment of mouse blastocysts on endometrial monolayers. Journal Reprod. Fert. 83, 149-158.

25. Springer, G.F. (1984) T and Tn, general carcinoma autoantigens. Science 224, 1198-1206.

26. Springer, G.F., Cheinsong-Popov, R., Schirrmacher, V., Desoi, P.R., Tegtmeyer, H. (1983) Proposed molecular basis of murine tumor cell-hepatocyte interaction. J. Biol. Chem. 258, 5702-5706.

27. Boring, C.C., Squires, T.S., Tong, T. (1991) Glycoconjugate expression during embryogenesis and its biological significance. CA-A Cancer J. Clinicians 41:19-36.

28. Kopf, A.W., Rigel, D.S., Friedman, R.J. (1982) The rising incidence and mortality rate of malignant melanoma. J. Dermatol Surg. Oncol. 8:760-761.

29. Cutler, S.J., Myers, M.H., Green, S.B. (1975) Trends in survival rates of patients with cancer. N. Engl. J. Med., 293:122-124.

30. Rigel, D.S., Kopf, A..W., Friedman, R.J. (1987) J.

Am. Acad. Dermatol. 17:1050-1053.

31. Clark, W.H. (1969) The histogenesis and biologic behavior of primary human malignant melanomas of the skin. Cancer Res. 29:705-727.

32. Breslow, A. (1970) Thickness, cross-sectional areas and depth of invasion in the prognosis of cutaneous melanoma. Ann Surg. 172:902-908.

33. Sohler, A.J. (1991) CA-A Cancer J. Clinicians 41:197-199.

34. Katano, M., Saxton, R.E., Tsuchida, T., et al., (1986) Human IgM monoclonal andti-GD2 antibody: Reactivity to a human melanoma xenograft. Jpn. J. Cancer Res. 775:584-594.

35. Cheresh, D.A., Reisfeld, R.A., Varki, A.P. (1984) O-acetylation of disialoganglioside GD3 by human melanoma cells creates a unique antigenic determinant. Science 225:844-846.

36. Thurin, J., Herlyn, M., Hindsgaul, O., Stromberg, N., Karlsson, A., Elder, D., Steplewski, Z., Koprowski, H. (1985) Proton NMR and fast-atom bombardment mass spectrometry analysis of the melamona-associated ganglioside 9-O-acetyl-GD3. J. Biol. Chem. 260:14556-14563.

37. Eggens, I., B. Fenderson, T. Toyokuni, B. Dean, M.Stroud and S.I. Hakomori (1989) Specific interaction between Le^x and Le^x determinants: A possible basis for cell recognition in preimplantation embroys and in embryonal carcinoma cells. J. Biol. Chem. 264:9476-9484.

38. Kojima, N., S.I. Hakomori. (1989) Specific interaction between gangliotriaosylceramide (Gg3) and sialosyllactosylceramide (GM3) as a basis for specific cellular recognition between lymphoma and melanoma cells. J. Biol. Chem. 264:20159-20162.

39. Glinsky, G.V. (1990) Immunoselective hypothesis of tumor progression. Role aberrant glycosylation, anticarbohydrate antibodies, extracellular glycomacromolecules and glycoamines. J. Tumor Marker Oncology. 5:206.

40. Glinsky, G.V. (1990) Glycoamines, aberrant glycosylation and cancer: a new approach to the understanding of molecular mechanism of malignancy. In: Molecular Oncology. Oncodevelopment proteins and clinical applications. XVIIIth meeting of the International Society for Oncodevelopmental Biology and Medicine. Abstract Book, Moscow, USSR, September 23-27, 1990, p.7.

41. Glinsky, G.V. (1992) Glycoamines: structural-functional characterization of a new class of human tumor markers. In: Serological Cancer Markers. Editor: S. Sell. The Humana Press, Totowa, NJ, Chapter 11, pp. 233-260.

42. Glinsky, G.V., Semyonova-Kobzar, R.A., Berezhnaya, N.M. (1990) Modification of cellular adhesion, metastasizing and immune response by glycoamines: implication in the pathogenetical role and potential therapeutic application in tumoral disease. J.Tumor Marker Oncology. 5:231.

43. Clausen, H., Hakomori, S.I. (1989) ABH and related histo-blood group antigens; Immunochemical differences in carrier isotypes and their destribution.

Vox Sang. 56, 1-20.

44. Lloyd, K.O. (1988) In: Altered glycosylation in tumor cells. Editors: Ch.L. Reading, S.-I. Hakomori; D.M. Markus, NY., Alan R. Liss, pp. 235-243.

45. Stein, R., Chen, S., Grossman, W., Goldenberg, D.M. (1989) Human lung carcinoma monoclonal antibody specific for the Thomsen-Friedenreich Antigen., Cancer Res. 49:32-37.

46. Fukuda, M., Spooncer, E. S., Oates, J.E., Dell, A., Klock, J.C. (1984) Structure of sialylated fucosyl lactosaminoglycan isolated from human granulocytes. J. Biol. Chem. 259:10925-10935

47. Symington, F.W., Hedges, D.L., Hakomori, S.-I. (1985) Glycolipid antigens of human polymorphonuclear neutrophils and the inducible HL-60 myeloid leukemia line. J. Immunol. 134:2498-2506.

48. Hakomori, S.-I. (1991) Possible functions of tumor-associated carbohydrate antigens. Current Opinion in Immunology 3:646-653.

49. Kojima, N., Hakomori, S. (1991) Cell adhesion, spreading and motility of GM3-expressing cells based on glycolipid-glycolipid interaction. J. Biol. Chem. 266:17552-17558.

50. Kojima, N., Hakomori, S. (1992) Synergistic effect of two cell recognition systems: glycosplingolipid-glycosphingolipid interaction and integrin receptor interaction with pericellular matrix protein. Glycobiology (in press)

51. Lawrence, M.B., Springer, T.A. (1991) Leukocytes roll on a selectin at phisiologic flow rates: Distinction from and prerequisite for adhesion through integrins. Cell 65:859-873.

52. Hakomori, S. (1981) Glycosphingolipids in cellular interaction, differentiation and oncogenesis. Annu. Rev. Biochem. 50:733-764.

53. Marchase, R.B. (1977) Biochemical investigations of retinotectal adhesive specificity. J. Cell Biol. 75:237-257.

54. Pierce, M. (1982) Quantification of ganglioside GM1 synthetase activity on intact chick neural retinal cells. J. Cell Biol. 93:76-81.

55. Obata, K., Oide, M., Handa, S. (1977) Effects of glycolipids on in vitro development of neuromuscular junction. Nature 266:369-371.

56. Fenderson, B.A., Zehavi, U., Hakomori, S. (1984) A multivalent lacto-N-fucopentaose III-lysyllysine conjugate decompacts preimplantation mouse embryos, while the free oligosaccharide is ineffective. J. Exp. Med. 160:1591-1596.

57. Bird, J.M., Kimber, S.J. (1984) Oligosaccharides containing fucose linked alpha(1-3) and alpha(1-4) to N-acetylglucosamine cause decompaction of mouse morulae. Dev. Biol. 104:449-460.

58. Yates, A.J., Thompson, D.K., Boesel, C.P., Albrightson, C., Hart, R.W., (1979) Lipid composition of human neural tumors. J. Lipid Res., 20:428-436

59. Dippold, W.G., Lloyd, K.O., Li, L.T.C., Ikeda, H., Oettgen, H.F., Old, L.J., (1980) Cell surface antigens of human malignant melanoma: definition of six antigenic systems with mouse monoclonal anti-

bodies. Proc. Natl. Acad. Sci (Wash.) 77:6114-6118.

60. Pukel, C.S., Lloyd, K.O., Travassos, L.R., Dippold, W.G., Oettgen, H.F., Old, L.J., (1982) GD3, a prominent ganglioside of human melanoma. J. Exp. Med. 155:1133-1147.

61. Schengrund, C.L., Repman, M.A., Shochat, S.J., (1985) Ganglioside composition of human neuro-blastomas. Cancer 56:2640-2646.

62. Wu, Z.L., Schwartz, E., Seeger, R., Ladisch, S., (1986) Expression of GD2 ganglioside by untreated primary human neuroblastomas. Cancer Res. 46:440-443.

63. Felding-Habermann, B.,ers, A., Dippold, W.G., Stallcup, W. B., Wiegandt, H., (1988) Melanoma-associated gangliosides in the fish Xiphophorus. Cancer Res. 48:3454-3460.

64. Tsuchida, T., Saxton, R.E., Morton, D.L., Irie, R.F. (1989) Gangliosides of human melanoma. Cancer 63:1166-1174.

65. Kono, K., Tsuchida, T., Kern, D.H., Irie, R.F., (1987) Ganglioside composition of human melanoma and response to anti-tumor treatment (Abstr.) Fed. Proc., 46:1057.

66. Tsuchida, T., Saxton, R.E., Morton, D.L., Irie, R.F. (1987) Gangliosides of human melanoma. J. Natl. Cancer Inst. 78:45-50.

67. Tsuchida, T., Otsuka, H., Nimura, M., et al. (1984) Biochemical study on gangliosides in neurofibromas and neurofibrosarcomas of Rickinghausen's disease. J. Dermatol. (Tokyo) 11:129-138.

68. Dippold, W. G., Dienes, H.P., Knuth, A., et al., (1985) Immunohistochemical localization of ganglioside GD3 in human malignant melanoma, epithelial tumors and normal tissues. Cancer Res. 45:3699-3705.

69. Natali, P., Bigotti, A., Caviliere, R., et al., (1985) Heterogeneous expression of melanoma-associated antigens and HLA antigens by primary and multiple metastatic lesions removed from patients with melanoma. Cancer Res. 45:2883-2889.

70. Suter, L., Brocker, E.B., Bruggen, J., Ruiter, D.J., et al., (1983) Heterogeneity of primary and metastatic human malignant melanoma as detected with monoclonal antibodies in cryostat sections of biopsies. Cancer Immunol. Immunother. 16:53-58.

71. Zehngebot, L.M., Alexander, M.A., Guerry, D.I., et al., (1983) Functional consequence of variation in melanoma antigen expression. Cancer Immunol. Immunother. 16:30-34.

72. Cillo, C., Mach, J.P., Schreyer, M., et al., (1984) Antigenic heterogeneity of clones and subclones from human melanoma cell lines demonstrated by a panel of monoclonal antibodies and flow microfluorometry analysis. Int. J. Cancer 34:11-20.

73. Portoukalian, J., Zwingelstein, G., Dore, J.F. (1979) Eur. J. Biochem. 94:19-23.

74. Houghton, A.N., Mintzer, D., Cordon-Cardo, C., Welt, S., Fliegel, B., Vadhan, S., Carswell, E., Melamed, M.R., Oettgen, H.F., Old, L.J., (1985) Mouse monoclonal IgG3 antibody detecting GD3

ganglioside: A phase I trial in patients with malignant melanoma. Proc. Natl. Acad. Sci. USA 82:1242-1246.

75. Dippold, W.G., Knuth, K.R., Meyer zum Buschenfelde, K.H. (1985) Inflammatory tumor response to monoclonal antibody infusion. Eur. J. Cancer Clin. Oncol. 21:907-912.

76. Houghton, A.N., Scheinberg, D.A Monoclonal antibodies: potential applications to the treatment of cancer. (1986) Semin. Oncol. 13:165-17980.

77. Cheresh, D.A., Varki, A.P., Varki, N.M., Stallcup, W.B., Levine, J.M., Reisfeld, R.A. (1984) A monoclonal antibody recognizes an O-acetylated sialic acid in a human melanoma-associated ganglioside. J. Biol. Chem. 259:7453-7459.

78. Ritter, G., Boosfeld, E., Markstein, E., Yu, R.K., Ren, Sh., Stallcup, W.B., Oettgen, H.F., Old, L.J., Livingston, Ph.O. (1990) Biochemical and serological characteristics of natural 9-O-acetyl GD3 from human melanoma and bovine buttermilk and chemically O-acetylated GD3. Cancer Res. 50:1403-1410.

79. Narasimhan, R., Murray, R.K. (1979) Neutral glycosphingolipids and gangliosides of human lung and lung tumors. Biochem. J. 179:199-211.

80. Irie, R.F., Sze, L.L., Saxton, R.E. (1982) Human antibody to OFA-I, a tumor antigen, produced in vitro by Epstein-Barr virus-transformed human B-lymphoid cell lines. Proc. Natl. Acad. Sci. USA, 79:5666-5670.

81. Schulz, G., Cheresh, D.A., Varki, N.M., et al., (1984) Detection of ganglioside GD2 in tumor tissues and sera of neuroblastoma patients. Cancer Res. 44:5914-5920.

82. Ariga, T., Takahashi, A., et al., (1980) Gangliosides and neutral glycolipids of human adrenal medulla. Biochim. Biophys. Acta. 618:480-485.

83. Rauvala, H. (1976) Isolation and partial characterization of human kidney gangliosides. Biochim. Biophys. Acta. 424:284-295.

84. Ladisch, S., Gillard, B., Wong, C., Ulsh, L., (1983) Shedding and immunoregulatory activity of YAC-1 lymphoma cell gangliosides. Cancer Res. 43:3308-3818.

85. Tai, T., Paulson, J.C., Cahan, L.D., Irie, R.F., (1983) Ganglioside GM2 as a human tumor antigen (OFA-I-1) Proc. Natl. Acad. Sci. (Wash.) 80:5392-5394.

86. Hirabayashi, Y., Hamaoka, A., Matsumoto, M., (1985) Syngeneic monoclonal antibody against melanoma antigen with interspecies cross-reactivity recognized GM3, a prominent ganglioside of B16 melanoma. J. Biol. Chem. 260:13328-13333.

87. Nilsson, O., Brezicka, F.T., Holmgren, J., Sorenson, S., Svnnerholm, L., Yngvason, F., Lindholm, L. (1986) .Detection of a ganglioside antigen associated with small-cell lung carcinomas, using monoclonal antibodies directed against fucosyl-GM1. Cancer Res. 46:1403-1407.

88. Herlyn, M., Rodeck, U., Koprowski, H. (1987) Shedding of tumor-associated antigens in vitro and in vivo. Adv. Cancer Res. 49:189-221.

89. Bernhard, H., Meyer zum Buschenfelde, K.-H.,

Dippold, W.G. (1989) Ganglioside GD3 shedding by human malignant melanoma cells. Int. J. Cancer 44:155-160.

90. Portoukalian, J., Zwingelstein, G., Abdul-Malak, N., Dore, J.F., (1978) Alteration of gangliosides in plasma and red cells of humans bearing melanoma tumors. Biochem. Biophys. Res. Commun. 85:916-920.

91. Schulz, G., Cheresh, D.A., Varki, N.M., Yu, A., Staffileno, L.K., Reisfeld, R.A. (1984) Detection of ganglioside GD2 in tumor tissues and sera of neuroblastoma patients. Cancer Res. 44:5914-5920.

92. Ladisch, S., Wu, Z.-L. (1985) Detection of tumor-associated ganglioside in plasma of patients with neuroblastoma. Lancet I:136-138.

93. Ladisch, S., Wu, Z., Feig, S., Ulsh, L., Schwartz, E., Floutsis, G., Wiley, F., Lenarsky, C., Seeger, R., (1987) Shedding of GD2 ganglioside by human neuroblastoma. Int. J. Cancer 39:73-76.

94. Yamanaka, T., Hirabayashi, T., Hirota, M., et al., (1987) Detection of gangliotriaose-series of glycosphingolipids in serum of cord blood and patients with neuroblastoma by a sensitive TLC/enzyme-immunostaining method. Biochim. Biophys. Acta. 920:181-184.

95. Tai, T., Cahan, L.D., Paulson, J.C., et al., (1984) Human monoclonal antibody against ganglioside GD2. Use in development of enzyme-linked immunosorbent assay for the monitoring of anti-GD2 in cancer patients. J. Natl. Cancer Inst. 73:627-633.

96. Ravindranaths, M.H., Paulson, J.C., Irie, R.F. (1988) Human melanoma antigen O-acetylated ganglioside GD3 is recognized by cancer antennarius lectin. J. Biol. Chem. 23:2079-2086.

97. Sela, B.-A., Iliopoulos, D., Guerry, D., Herlyn, D and Koprowski, H. (1989) Levels of disialogangliosides in sera of melanoma patients monitored by sensitive thin-layer chromatography and immunostaining. J. Natl. Cancer Inst. 81:1489-1492.

98. Ross, A.H., Herlyn, M., Ernst, C.S., et al., (1984) Immunoassay for melanoma-associated proteoglycan in the sera of patients using monoclonal and polyclonal antibodies. Cancer Res., 44:4642-4647.

99. Koprowski, H., Herlyn, M. Steplewski, Z., et al., (1981) Specific antigen in serum of patients with colon carcinoma. Science 212:53-55.

100. Bast, R.C., Jr., Klug, T.L., St. John, E., et al., (1983) A radioimmunoassay using a monoclonal antibody to monitor the course of epithelial ovarian cancer. N Engl. J. Med. 309:883-887.

101. Higashi, H., Hirabayashi, Y., Hirota, M., et al., (1987) Detection of ganglioside GM2 in sera and tumor tissues of hepatoma patients. Jpn. J. Cancer Res. 78:1309-1313.

102. Fukushi, Y., Kannagi, R., Hakomori, S., Shepard, T., Kulander, B.G., Singer, J.W., (1985) Location and distribution of difucoganglioside in normal and tumor tissues defined by its monoclonal antibody FH6. Cancer Res. 45:3711-3717.

103. Ladisch, S., Ulsh, L., Gillard, B., Wong, C. (1984) Modulation of the immune response by gangliosides: Inhibition of adherent monocyte accessory function in vitro. J. Clin. Invest. 74: 2074-2081.

104. Gonwa, T.A., Westrick, M.A., Macher, B.A., (1984) Inhibition of mitogen-and antigen-induced lymphocyte activation by human leukemia cell gangliosides. Cancer Res. 44:3467-3470.

105. Ando, I., Hoon, D.S.B., Suzuki, Y., Saxton, R.E., Golub, S.H., Irie, R.F., (1987) Ganglioside GM2 on the K562 cell line is recognized as a target structure by human natural killer cells. Int. J. Cancer 40:12-17.

106. Offner, H., Thieme, T., Vandenbark, A.A., (1987) Gangliosides induce selective modulation of CD4 from helper T lymphocytes. J. Immunol. 139:3295-3305.

107. Merritt, W.D., Bailey, J.M., Pluznik, D.H., (1984) Inhibition of interleukin-2-dependent cytotoxic T-lymphocyte growth by gangliosides. Cell Immunol. 89:1-10.

108. Robb, R.J. (1986) The suppressive effect of gangliosides upon IL-2-dependent proliferation as a function of inhibition of IL-2-receptor association. J. Immunol. 136:971-976.

109. Yogeeswaran, G., Hakomori, S. (1975) . Cell-contact-dependent ganglioside changes in mouse 3T3 fibroblasts and a suppressed sialidase activity on cell contact. Biochemistry 14:2151-2156.

110. Hakomori, S., Kannagi, R., (1983) Glycosphingolipids as tumor-associated and differentiation markers. J. Natl. Cancer Inst. 71:231-251.

111. Felding-Habermann, B., Jennemann, R., Schmitt, J., Wiegandt, H., (1986) Glycosphingolipid biosynthesis in early chick embryos. Europ. J. Biochem. 160: 651-658.

112. Sariola, H., Aufderheide, E., Bernhard, H., Henke-Fahle, S., Dippold, W., Ekblom, P. (1988) Antibodies to cell surface ganglioside GD3 perturb inductive epithelial-mesenchymal interactions. Cell 54:235-245.

113. Morgan, J.I., Seifert, W., (1979) Growth factors and gangliosides: a possible new perspective in neuronal growth control. J. Supramolecular Structure. 10:111-124.

114. Bremer, E.G., Hakomori, S., Bowen-Pope, D.F., Raines, E., Ross, R., (1984) Ganglioside-mediated modulation of cell growth, growth factor binding and receptor phosphorylation. J. Biol. Chem. 259:6818-6825.

115. Bremer, E.G., Schlessinger, J., Hakomori, S., (1986) Ganglioside-mediated modulation of cell growth. J. Biol. Chem. 261:2434-2440.

116. Katoh-Semba, R., Facci, L., Skaper, S.D., Varon, S., (1986) Gangliosides stimulate astroglial cell proliferation in the absence of serum. J. Cell Physiol. 126:147-153.

117. Nakakuma, H., Sanai, Y., Shiroki, K., Nagai, Y.(1984) Gene-related expression of glycolipids: appearence of GD3 gangliosides in rat cells on transfection with transforming gene E1 of human adeno-virus type 12 DNA and its transcriptional subunits. J. Biochem. 96:1471-1480.

118. Carubia, J.M., Yu, R.K., Macala, L.J., Kirkwood, J.M., Varga, J.M., (1984) Gangliosides of normal and neoplastic human melanocytes. Biochem. Biophys. Res. Comm., 120:500-504.

119. Albino, A.P., Houghton, A.N., Eisinger, M., Lee, J.S., Kantor, R.R. S., Oliff, A.I., Old, L.J. (1986) Class II histocompatibility antigen expression in human melanocytes transformed by Harvey murine sarcoma virus (Ha-MsV) and Kirsten MSV retrovirus. J. Exp. Med. 164:1710-1722.

120. Ladisch, S., Kitada, S., Hays, E.F., (1987a) Gangliosides shed by tumor cells enhance tumor formation in mice. J. Clin. Invest. 79:1879-1882.

121. Alessandri, G., Filippeschi, S., Sinibaldi, P,., Mornet, F., Passera, P., Spreafico, F., Cappa, P.M., Gullino, P.M. (1987) Influence of gangliosides on primary and metastatic neoplastic growth in human and murine cell. Cancer Res. 47:4243-4247.

122. Cheresh, D.A., Harper, J.R., Schulz, G., Reisfeld, R.A., (1984) Localization of the gangliosides GD2 and GD3 in adhesion plaques and on the surface of human melanoma cells. Proc. Natl. Acad. Sci (Wash.) 81:5767-5771.

123. Okada, Y., Mugnai, G., Bremer, E.G., Hakomori, S., (1984) Glycosphingolipids in detergent-insoluble substrate attachment matrix (DISAM) prepared from substrate attachment material (SAM) Exp. Cell Res. 155:448-456.

124. Cheresh, D.A., Klier, F.G. (1986) Disialoganglioside GD2 distributes preferentially into substrate-associated microprocesses in human melanoma cells during their attachment to fibronectin. J. Cell Biol. 102:1887-1897.

125. Tsuchida, T., Saxton, R.D., Irie, R.F., (1987) Gangliosides of human melanoma: GM2 and tumorigenicity. J. Natl. Cancer Inst. 78:55-60.

126. Allen, H.J., Karakousis, C., Piver, M.S., Gamarra, M., Nava, H., Forsyth, B., Matecki, B., Jazayeri, A., Sucato, D., Kisailus, E., DiCioccio, R. (1987) Galactoside-binding lectin in human tissues. Tumor Biol. 8:218-229.

127. Carding, S.R., Thorpe, S.J., Thorpe, R., Feizi, T. (1983) Transformation and growth related changes in levels of nuclear and cytoplasmic proteins antigenically related to mammalian beta-galactoside-binding lectin. Biochem. Biophys. Res. Commun., 127:680-686.

128. Gabius, H-J., Engelhardt, R., Cramer, F. (1986) Endogenous tumor lectins: overview and perspectives. Anticancer Res. 6:573-578.

129. Gitt, M.A., Barondes, S.H., (1986) Evidence that a human soluble beta-galactoside-binding lectin is encoded by a family of genes. Proc. Natl. Acad. Sci. USA, 83:7603-7607.

130. Raz, A., Meromsky, L., Carmi, P., Karakash, R., Lotan, D., Lotan, R. (1984) Monoclonal antibodies to endogenous galactose-specific tumor cell lectins. EMBO J. 3:2979-2983.

131. Raz, A., Meromsky, L., Lotan, R. (1986) Differential

expression of endogenous lectins on the surface of nontumorigenic, tumorigenic, and metastatic cells. Cancer Res. 46:3667-3672.

132. Raz, A., Avivi, A., Pazarini, G., Garmi, P. (1987) Cloning and expression of cDNA for two endogenous UV-2237 fibrosarcoma lectin genes. Exp. Cell Res. 173:109-116.

133. Raz, A., Meromsky, L., Zvibel, I., Lotan, R. (1987) Transformation-related changes in the expression of endogenous cell lectins. Int. J. Cancer 39:353-360.

134. Raz,A, Carmi. P., Pazarini, G., (1988) Expression of two different endogenous galactoside-binding lectins sharing sequence homology. Cancer Res. 48:645-649.

135. Glinsky, G.V. (1992) Glycodeterminants of melanoma cell adhesion. A model for antimetastatic drugs design. Crit. Rev. Onc. Hemat. (accepted)

136. Glinsky, G.V. (1989) Glycoamines: Biochemistry of a new class of humoral tumor markers. J. Tumor Marker Oncology. 4:193-221.

137. Glinsky, G.V., Nikolaev, V.G., Ivanova, A.B., Shemchuk, A.S., Lisetsky, V.A. (1980) "Middle molecules" in cancer patients blood plasma. Experimental Oncology. V.2, N 1, p. 68-71.

138. Clinsky, G.V., Linetsky, M.D., Ivanova, A.B., Osadchaya, L.P., Semyonova-Kobzar, R.A., Surmilo, N.I. (1990) Chemical, biochemical, spectroscopic and biological identification of structural-functional determinants of glycoamines (aminoglycoconjugates) J. Tumor Marker Oncology 5:250.

139. Glinsky, G.V., Ogorodniychuk, A.S., Shilin, V.V., Linetsky, M.D., Livenstov, V.V., Sidorenko, M.V., Surmilo, N.I., Kukhar, V.P. (1990) Structural analysis of synthetic structural analogs of glycoamines: contribution into elucidation of the structure and biogenesis of natural aminoglycoconjugates. J. Tumor Marker Oncology 5:250.

140. Oguchi, H., Toyokuni, T., Dean, B., Ito, H., Otsuji, E., Jones, V.L., Sadozai, K.K., Hakomori, S., (1990) Effect of lactose derivatives on metastatic potential of B16 melanoma cells. Cancer Communication 2:311-316.

CHAPTER 8

AIDS PATHOGENESIS:
A KEY FUNCTION OF BGA-RELATED GLYCODETERMINANTS

The concept that HIV is the sole cause of AIDS and is directly responsible for the depletion of CD4-bearing T cells is currently being questioned by many researchers for several reasons:[1] low numbers of T cells are infected with HIV in AIDS,[2] and immunosuppression occurs prior to depletion of CD4 cells;[3] immunity against HIV does not protect against AIDS; AIDS patients have high quantities of anti-HIV antibodies in their sera[4] and generate a strong anti-HIV cytotoxic T-cell response[5] but are not protected. An alternative possibility being considered by several groups of researchers is that HIV induces a deleterious immune response that attacks the immune system itself.[6-15] Human immunodeficiency virus type 1 (HIV-1) is tropic and cytopathic for human T lymphocytes bearing the CD4 receptor.[16,17] Macrophages, which also express CD4 on the cell surface, have also been shown to play a major role in the propagation and pathogenesis of HIV-1 infection.[18-21] The most important implication of the infectivity of monocytes with HIV is the possibility that the monocyte serves as the major reservoir for HIV in the body.[22] Unlike the T4 lymphocyte, the monocyte is relatively refractory to the cytopathic effects of HIV so that not only can the virus survive in this cell but it can be transported to various organs in the body such as the lung and the brain. The monocyte precursors may be noncytopathologically infected and capable of secreting virus upon appropriate induction by certain cytokines.[23] HIV-1 preferentially infects cells expressing CD4 on their surface,[24-26] however, three independent lines of evidence support the idea that HIV-1 tropism may not be limited to CD 4-positive cells and that there is a non-CD4 pathway of HIV infection.

HIV IS CAPABLE OF INFECTING NOT ONLY CD4-BEARING CELLS

Many human cell lines which were not known to be CD4 positive have recently been found to be susceptible to HIV-1 infection in vitro.[27] These include cells derived from colorectal carcinoma,[28] rhabdomyosarcoma,[29] bone marrow CD 34+ precursor cells,[30] glioma,[29,31-34] neuroblastoma[35] and hepatoma.[27] Several reports have also shown that HIV-1 is capable of infecting certain nonlymphoid tissues or cells in vivo, including colon, rectum, duodenum, cervix, retina, brain, and megakaryocytes.[36-40] These in vitro and in vivo findings suggest that HIV-1 tropism may not be limited to CD4-positive cells, and that there is a non-CD4 pathway by which the virus can enter cells, such as those of glial,[29,31-34] fibroblastoid[41] and neuronal[35] origin and rhabdomyosarcoma.[29]

Unintegrated HIV-1 DNA Accumulation in Target Cells is not Dependent on CD4 Receptor Number

Infection with HIV-1 may lead either to destruction of the host cell or to a persistent, noncytopathic infection. The outcome of infecion with HIV-1 has been correlated with the amount of unintegrated viral DNA present in the host cell.[42] However, significant amounts of unintegrated viral DNA are not generally observed in persistent HIV-1 infections. This has been attributed to protection from reinfection due to the loss or down-modulation of the CD4 receptor, a feature of most chronically infected cell lines.[43-45] As the level of unintegrated retroviral DNA in vitro has been correlated with cytopathicity, so the level of unintegrated retroviral DNA in vivo has been correlated with disease.[46,47] Such DNA accumulates to high levels in the cells as a result of second-round superinfection and may kill the cell directly or indirectly through an unknown mechanism.[42,48] Unintegrated HIV-1 DNA is unstable, originates from a continuous process of reinfection, and the number of CD4 receptors is not rate limiting for reinfection.[49]

HIV Infection may be Mediated not only by GP 120

The density of glycoprotein knobs covering the surface of HIV particles decreases with time. This spontaneous shedding occurs because the gp 120 surface and gp 41 transmembrane glycoproteins are associated by noncovalent interactions.[50,51] After observing this gp 120 loss by electron microscopy, Gelderblom et al[52,53] suggested that knob density might influence the biological properties, e.g., infection and blocking, of HIV. Soluble CD4 and other gp 120 blocking agents and neutralization domains have been used for blocking of HIV infection. Examples of such blockers include CD4 fragments,[54,55] CD4-immunoglobulin conjugates[56-58] and

some vaccine-induced immunoglobulins.[59] The other blocking domains on gp 120 include the third variable region (V3 loop),[60-65] conformational epitopes[66-69] and conserved sequences.[70-72] However, the blocking of HIV infection by lectins and monoclonal anti-carbohydrate antibodies has been recently reported.[73-75]

Carbohydrates are Involved in HIV Infection

The envelope glycoproteins of human immunodeficiency virus type 1 (HIV-1) are highly glycosylated molecules, N-linked glycans representing one-half of their molecular weight.[76,77] Therefore, carbohydrates are likely to be prominent structures on the surface of HIV-1, but their precise function is not understood. This role is likely to be important because one-half of the asparagine residues of the env sequence are involved in its approximately 30 glycosylation sites and because these sites are rather conserved when one compares different HIV-1 isolates or HIV-1 with HIV-2 and simian immunodeficiency virus.[78,79] Only N-linked glycans have been found on gp 120.[80] The core of N-linked glycosylation of gp 120 of recombinant HIV[81] as well as of gp120 of HIV from human T-cells (H9 cells) has been investigated.[82]

Virus produced in the presence of glucosidase inhibitors, which result in aberrant glycosylation, displays markedly reduced infectivity and cytopathogenicity.[83] It has also been suggested that carbohydrates at the surface of the mature molecule may play a direct role in the interaction of gp 120 with CD4[73,84] and enzymatic removal of glycans from envelope glycoproteins results in the reduction either of binding to CD4 or of viral infectivity for CD4+ lymphoid cells.[85] Matthews et al also reported that enzymatic deglycosylation of soluble viral gp 120 from the supernatant of HIV-1-infected H9 cells markedly reduced its capacity to bind to membrane CD4 and to inhibit syncytium formation.[77] Altered glycosylation in host cells associated with viral infection has been reported.[86,87] Aberrant glycosylation induced by cytomegalovirus or by HIV causes formation of neoantigens which are absent in the original host cells. By using monoclonal antibodies (mAb) which define oligosaccharide epitopes, the appearance of LeX and LeY antigens after viral infection has been detected.[88,89]

Thus, several studies have indicated the involvement of the carbohydrate part of HIV in in vitro infection. Inhibition of the early steps in Golgi glycosylation in infected cells reduces the infectivity of the virus produced,[83,90] and lectins block syncytium formation probably by a specific interaction with gp120 glycans of infected cells, and also neutralize infectivity of cell-free virus.[73,74] However, variation in N-glycosylation of target T4 cells does not seem to influence susceptibility to HIV infection.[90] The binding site on gp120 for the T4 receptor seems to be located in a nonlinear C-terminal part of the molecule.[91] Whether glycan(s) participates directly in virus binding is not clear. Thus, inhibitory lectins may bind to glycans adjacent to the binding site and thereby sterically interfere with T4 gp120 binding, as has been found for neutralizing antibodies.[92,93] Carbohydrate residues on the HIV envelope glycoprotein gp120 have been implicated in adhesion to CD4, initiating HIV infection of lymphocytes. Thus, deglycosylation of gp120 abrogates binding to the CD4 receptor,[77] and inhibition of Golgi-mediated glycosylation in HIV-infected cells reduces the infectivity of the HIV produced.[83,90] A similar antiviral effect of glycosylation inhibition has been found in vivo in a murine system in which Rauscher leukemia virus was used.[94]

BGA-Related Glycoepitopes are Directly Involved in HIV Infection

Hansen et al have reported that mAb BM1, directed to LeY previously found to be expressed on HIV-infected lymphocytes,[74,75] inhibited formation of syncytia between infected and uninfected cells and also the infectivity of cell-free virus.[75] This inhibition was observed only on preincubation of mAb with virus, but not of mAb with cells before inoculation. Furthermore, BM1 precipitated gp120 in an ^{125}I-labeled HIV lysate. Two mAbs, AH21 (IgM) and AH 16 (IgG3), with specificity for blood group A type 1 chain antigen,[96] also showed inhibition, whereas mAbs directed to other A-antigen types did not.[75] The finding that mAbs TKH2 and B72.3, directed to the sialyl-Tn antigen, inhibited infection and syncytium formation and also precipitated gp120, suggests that O-linked glycans may also exist on the HIV envelope[75] as has been found in other enveloped viruses.[97] The glycosylation of viral proteins is believed to be performed by the host cell gene-encoded glycosyltransferases. Therefore, it is expected that the cell and proteins of viruses propagated therein show similar glycosylation patterns. The three carbohydrate neutralization epitopes (LeY, A1 and sialyl-Tn) found to be expressed in chronically HIV-infected lymphocytes[75] which corresponds well with previous studies of LeY expression,[88,95] and for at least one epitope (sialyl-Tn), increased expression after HIV infection has been found.[75]

BGA-Related Glycoantigens are Transitory Expressed on Human Tissues During Development

Glycoantigens undergo dramatic changes during development, cell differentiation and maturation.[98-100] We will only consider one example that clearly shows the appearance, disappearance and reappearance of BGA-related glycoantigens on the cell surface at definite stages of cell differentiation during embryogenesis, organogenesis, tissue formation and maturation. Cancer cells frequently express carbohydrate embryonic antigens since they undergo a retrodifferentiation process during the course of malignant transformation.[101] SSEA-1 (stage-specific embryonic antigen-1) first described by Solter and Knowles in 1978[102] is one of the embryonic antigens which is specifically expressed on the murine preimplantation embryos at the morula stage. The SSEA-1 antigen is described chemically as a carbohydrate antigen carrying LeX-hapten and i-antigenic structures.[103,104] By using specific monoclonal antibodies, it was shown that SSEA-1 and its modified form of the antigen, including the sialylated form (sialylated LeX or sialylated LeX-i) and fucosylated forms (LeY and poly LeX), are frequently accumulated in human cancer of various origins.[105-109] Since SSEA-1 was first described as an embryonic antigen and subsequently found in various human cancers, the antigen has sometimes been regarded as an oncofetal or oncodevelopmental antigen.[110-114]

The localization of three carbohydrate antigens, LeX, LeY and sialylated LeX-, which are closely related to stage-specific embyryonic antigen 1 in the lung of developing human embryos, was investigated using specific monoclonal antibodies.[101] The three antigens all have a physiological significance as stage-specific developmental antigens of the human lung; those antigens were specifically present in the bud cells at each important step of the morphogenesis of the human lung, such as cells in the lung buds, bronchial buds and terminal buds for the formation of the alveolus, and cells differentiating into bronchial gland cells. The three antigens gradually disappear in the later stage of development along with the maturation process of the lung. Stage-specific embryonic antigen 1 and related antigens are known to be associated with various human cancers including lung cancer.

Miyake et al[101] suggest that the expression of these antigens in lung cancer cells is the result of the retrodifferentiation of the cancer cells to the stages of immature embryonic lung cells. It was known that SSEA-1 antigens reappear in a certain lineage of cells after implantation during the course of morphogenesis of certain organs such as brain and urogenital organs,[113,115] and the stage specific expression of LeX and related antigens in developing lungs of human embryos was shown.[101] One of the types of expression of the stage-specific antigen is characterized by strict stage specificity and disappearance of the antigen in mature cells. This expression of SSEA-1-related antigen during the course of organogenesis has been regarded as the second wave of expression of those antigens,[101] assuming the appearance of these antigens in the preimplantation embryos is classically described[102] as the first wave of the expression. The third type of the expression of stage-specific antigen is characterized by this persistency of the antigen after differentiation. This expression of the SSEA-1 related antigens has been regarded as the third and the latest wave of the expression of these antigens in intrauterine life.[101]

CONCLUSION: AN IMPLICATION IN AIDS PATHOGENESIS

Thus, a large body of evidence has accumulated that indicates the existence of a CD4+-independent pathway of HIV infection and AIDS pathogenesis. Like many other viruses and bacteria,[116] HIV may use the glycodeterminants and/or carbohydrate recognition structures on cell surface as receptors. Particularly important is that those glycodeterminants have been identified as a BGA-related glycoepitopes (see above). Like embryogenesis, tissue regeneration, repair and remodeling could be accompanied by expression of BGA-related glycoepitopes at definite stages of cell differentiation. At a particular stage of differentiation, cells that express BGA-related glycoepitopes and/or BGA-recognition structures become targets of HIV-infection. During the course of HIV infection and AIDS development, HIV could use a receptor-driving mechanism, e.g., change the predominant protein-protein type of ligand-receptor interaction at the first stage of infection to the carbohydrate-protein or carbohydrate-carbohydrate types of virus ligand-target cell receptor interaction at the late stage of infection. HIV could use this mechanism to dramatically increase the spectrum of target cells and tissues and generalization of infection throughout the host body. The molecular basis for this hypothesis is provided by data that show similarities between class II MHC and HIV proteins.

Sequence homologies have been reported between a conserved region of gp 120 and MHC class II[117] and also between Nef of HIV and MHC class II.[118] Ziegler and Stites[15] and Andrieu et al[6] have suggested that gp 120 could resemble class II MHC and the immune response to gp 120 could cross-react with class II MHC since the gp 120 envelope glycoprotein is complementary to CD4[91] and CD4 is complementary to class II MHC.[119]

Recombinant gp 120 can block the interaction of CD4 with MHC class II[120-122] as can antibodies that recognize the gp 120 binding site of CD4.[123] Thus, the binding sites for gp 120 and class II MHC overlap in the first immunoglobulin-like domain of CD4. This conclusion has been supported by analysis of the effects of mutations in CD4 on the interactions of CD4 with gp 120 and MHC class II: while the binding sites are separable because some mutations affect the gp 120 binding site but not the class II MHC binding site and vice versa, there are also some mutations that affect both interactions.[124,125] A serological similarity between gp 120 of HIV and class II MHC has been reported.[126,127] Because of these structural similarities, it is very possible that HIV glycoproteins become glycosylated identically (or very similar) to the host cell glycoproteins, in particular class II MHC molecules, after HIV infection of primarily target cells. Consequently, HIV glycoproteins could have (after glycosylation by host cell glycosyltransferases), identically to host cells, unique patterns of BGA-related glycoepitopes and subsequently HIV could use these highly specific carbohydrate structural units for expansion of infection. The idea that MHC molecules are carriers for BGA-related glycoepitopes provides a molecular basis for a structural link between ABO and HLA systems.

According to this concept, cancer and AIDS have common, critical stages of disease pathogenesis that involve BGA-related glycoepitopes as key structural determinants, and represent the mechanism of generalization of the pathological process throughout the body.[128] In the case of AIDS, the effector part of this mechanism is presented by HIV glycoproteins; in the case of cancer, it is presented by cancer cells and serum glycomacromolecules. Since the postulated critical role of BGA-related glycoepitopes in cell differentiation during tissue remodeling, repair, regeneration and immunogenesis, homeostasis will be affected and destroyed in both cases—AIDS and cancer—and immunosuppression and immunodeficiency will ensue. The lysis of infected T cells and macrophages may be an important aspect for host control

of HIV infection. However, the same immune effector mechanism may also lead to the destruction of noninfected CD4+ T cells that have taken up free gp120.[14] A subset of CD4+ gp120-specific T cell clones manifests cytolytic activity and lyses uninfected autologous CD4+ Ia + T cells in the presence of gp120 in a process that is strictly dependent upon CD4-mediated uptake of gp120 by target T cells. Since gp120 is shed from HIV-infected cells in vivo (see above), this CD4-dependent autocytolytic mechanism may contribute to the profound depletion of CD4+ cells in AIDS.[14] Correspondingly, carbohydrate-mediated gp120 uptake by uninfected host cells could initiate a CD4-independent BGA-related glycoantigen-mediated gp120-specific autoimmune cytolytic mechanism that may contribute to AIDS pathogenesis through autoimmune destruction of a broad range of histogenetically different uninfected host cells that express BGA-related glycodeterminants and/or corresponding carbohydrate-recognition structures.

The selectins, major adhesion proteins of circulating cells, e.g., leukocytes, platelets, and endothelium, play a key role in adhesive interactions that direct neutrophil localization in inflammation, lymphocyte localization and migration and determine lymphocyte homing to different lymphoid organs.[129-131] Carbohydrate ligands for selectins have recently been discovered: sialosyl-LeX, sialosyl-LeA, and LeX have been identified as recognition structures for ELAM -1 and CD62.[132-135] One of the selectins, a molecule termed endothelial cell-adhesion molecule -1 (ELAM -1), was originally implicated in the binding of blood neutrophils and monocyte-like cells to inflammed endothelium.[136] Picker, et al[137] and Shimizu et al[138] showed that ELAM-1 is one of the receptors on endothelium for a specific subset of circulating T lymphocytes: only memory-type T cells (CD4+ T cells) bind ELAM-1. Thus, ELAM-1 molecules may be responsible for specific distribution and migration patterns of memory T cells.[139] Their accumulation in inflammed lesions and recirculation through uninflammed tissues including skin.[140]

On the other hand, monoclonal antibodies against three different N- and O-linked carbohydrate epitopes (LeY, A1 and sialyl-Tn) were able to block HIV infection by cell-free virus in vitro as well as inhibit syncytium formation,[75] and the appearence on the host cells of LeX and LeY antigens after viral infection has been detected.[88,89] The mAbs define carbohydrate structures expressed by the viral envelope glycoprotein gp 120 since mAbs were shown to precipitate ^{125}I-labeled gp120.[75] The density of glycoprotein knobs covering the surface of HIV particles decreases with time.[52,53] This spontaneous shedding occurs because the gp120 surface and gp41 transmembrane glycoproteins are associated by noncovalent interactions.[50,51]

Taken together those findings suggest that: a) endothelial selectins, particularly ELAM-1, may serve as receptors for a HIV infected subset of T lymphocytes that may determine the specific recirculation pathway of HIV infected cells and spread of infection; b) circulating gp120 may affect physiologic trafficking of memory T cells through blocking of their interaction with ELAM-1 and therefore, may inhibit the function of CD4+ T cells during AIDS, without cell infection and/or depletion. One of the intriguing conclusions from this speculation is that prospective carbohydrate-based leukocyte-endothelial cell adhesion inhibitor(s) designed as an antiinflammatory agent[130] may well be an effective anti-AIDS drug.[141]

References

1. Hoffmann, G.W., Kion, A.T., Grant, M.D. (1991) An idiotypic network model of AIDS immunopathogenesis. Proc. Natl. Acad. Sci. USA 88:3060-3064.

2. Schnittman, S.M., Psallidopoulos, M.C., Lane, H.C., Thompson, L., Baseler, M., Masari, F., Fox, C.H. Salzman, N.P., Fauci, A. (1989) The reservoir for HIV-1 in human peripheral blood is a T cell that maintains expression of CD4. Science 245:305-308.

3. Giorgi, J.V., Fahey, F.L., Smith, D.C., Hultin, L.E., Cheng, H.L., Mitsuyasu, R.T., Detels, R. (1987) Early effects of HIV on CD4 lymphocytes in vivo. J. Immunol. 138:3725-3730.

4. Karpatkin, S., Nardi, M., Lenette, E.T., Byrne, B., Poiesz, B. (1988) Anti-human immunodeficiency virus type 1 antibody complexes on platelets of seropositive thrombocytopenic homosexuals and narcotic addicts. Proc. Natl. Acad. Sci. USA 85:9763-9767.

5. Hoffenbach, A., Langlade-Demoyen, P, Dadaglio, G., Vilmer, E., Michel, F., Mayaud, C., Autran, B., Plata,F. (1989) Unusually high frequencies of HIV-specific cytotoxic T lymphocytes in humans. J. Immunol. 142:452-462.

6. Andrieu, J.M., Even, P., Venet, A. (1986) AIDS and related syndromes as a viral-induced autoimmune disease of the immune system—an anti-MHC II disorder: Therapeutic implications. AIDS Res. 2:163-174.

7. Calabrese, L.H. (1988) Autoimmune manifestations of human immunodeficiency virus (HIV) infection. Clin. Lab. Med. 8:269.

8. Habeshaw, J.A., Dalgleish, A. (1989) The relevance of HIV env/CD4 interactions to the pathogenesis of acquired immune deficiency syndrome. J. AIDS 2:457-468.

9. Hoffmann,G.W. (1988) In: The Semiotics of Cellular Communication in the Immune System. eds. Sercarz, E.E., Celada, F., Mitchison, N.A., Tada, T. (Springer, New York), pp. 257-271.

10. Hoffmann, G.W., Grant, M.D. (1989) Lect. Notes Biomath. 83:386-401.

11. Lanzavecchia, A. (1989) Harming and protecting responses to HIV. Res. Immunol. 140:99-103.

12. Martinez-A., C., Marcos, M.A.R., de la Hera, A., Marquez, C., Alonso, J.M., Toribio, M.L., Coutinho, A. (1988) Immunological consequences of HIV infection: advantage of being low responder casts doubts on vaccine development. Lancet I,454-457.

13. Shearer, G.M. (1986) Mt. Sinai J. Med. N.Y. 53:609-615.

14. Siliciano, R.F., Lawton, T.C., Knall, C., Karr, R.W., Bermann, P., Gregory, T., Reinherz, E.L. (1988) Analysis of host-virus interactions in AIDS with anti-gp120 T cell clones: Effect of HIV sequence variation and a mechanism for CD4+ cell depletion. Cell 54:561-575.

15. Ziegler, J.L., Stites, D.P. (1986) Hypothesis: AIDS is an autoimmune disease at the immune system and triggered by a lymphotropic retrovirus. Clin. Immunol. Immunopathol. 41:305-313.

16. Klatzmann, D., Barre-Sinoussi, F., Nugeryre, M.T., et al. Chermann and L. Montagnier. (1984) Selective tropism of lymphadenopathy-associated virus (LAV) for helper-inducer T-lymphocytes. Science 225:59-63.

17. Popovic, M., Sarngadharan, M.G., Read, E., R.C. Gallo, R.C. (1984) Detection, isolation and continuous production of cytopathic retroviruses (HTLV-III) from patients with AIDS and pre-AIDS. Science 224:497-500.

18 Gartner, S., Markovits, P., Markovitz, D.M., et al. (1986) The role of mononuclear phagocytes in HTLV-III/LAV infection. Science 233:215-219.

19. Ho, D.D., Rota, T.R., Hirsch, M.S. (1986) Infection of monocyte/macrophages by human T-lymphotropic virus type III. J. Clin. Invest. 77:1712-1715.

20. Nicholson, J.K.A., Gross, G.D., Callaway, C.S., McDougal, J.S. (1986) In vitro infection of human

monocytes with human T-lymphotropic virus type III/lymphadenopathy-associated viruses (HTLV-III/LAV) J. Immunol. 137:323-329.

21. Salahuddin, S.Z., Rose, R.M., Groopman, J.E., Markham, P.D., Gallo, R.C. (1986) Human T-lymphotropic virus type III infection of human alveolar macrophages. Blood. 68:281-284.

22. Fauci, A.S. (1988) The human immunodeficiency virus: Infectivity and mechanisms of pathogenesis. Science. 239:617-622.

23. Folks, T.M., Justement, J., Kinter, A., Dinarello, C.A., Fauci, A.S., (1987) Cytokine-induced expression of HIV-1 in a chronically infected promonocyte cell line. Science. 238:800.

24. Dalgleish, A.G., P.C.L. Beverley, P.R. Clapham, D.H. Crawford, M.F. Greaves and R.A. Weiss. 91984) The CD4 (T4) antigen is an essential component of the receptor for the AIDS retrovirus. Nature 312:763-767.

25. Klatzmann, D., Champagne, E., Chamaret, S., Gruest, J., Guetard, D., Hercend, T., Gluckman, J.C., Montagnier, L. (1984) T-lymphocyte T4 molecules behave as the receptor for human retrovirus LAV. Nature 312:767-768.

26. Maddon, P.J., Dalgleish, A.G., McDougal, J.S., Clapham, P.R., Weiss, R.A., Axel, R. (1986) The T4 gene encodes the AIDS virus receptor and is expressed in the immune system and the brain. Cell 47:333-348.

27. Cao, Y., Friedman-Kien, A.E., Huang, Y., Li, X.L., Mirabile, M., Moudgil, T., Zucker-Franklin, D., Ho, D.D. (1990) CD4-independent, productive human immunodeficiency virus type 1 infection of hepatoma cell lines in vitro. J. Virol. 64:2553-2559.

28. Adachi, A., Koenig, S., Gendelman, H.E., Daugherty, D., Gattoni-Celli, S., Fauci, A.S., Martin, M.A. (1987) Productive, persistent infection of human colorectal cell lines with human immunodeficiency virus. J. Virol. 61:209-213.

29. Clapham, P., Weber, J.N., Whitby, D., McKintosh, K., Dalgleish, A.G., Maddon, P.J., Deen, K.C., Sweet, R.W., Weiss, R.A. (1989) Soluble CD4 blocks the infectivity of diverse strains of HIV and SIV for T cells and monocytes but not for brain and muscle cells. Nature 337:368-370.

30. Folks, T.M., S.W. Kessler, J.M. Orenstein, J.S. Justement, E.S. Jaffe and A.S. Fauci. (1988) Infection and replication of HIV-1 in purified progenitor cells of normal human bone marrow. Science. 242:919-922.

31. Cheng-Mayer, C., Rutka, J.T., Rosenblum, M.L., McHugh, T., Stites, D.P., Levy, J.A. (1987) Human immunodeficiency virus can productively infect cultured human glial cells. Proc. Natl. Acad. Sci. USA 84:3526-3530.

32. Chiodi, F., Fuerstenberg, S., Gidlund, M., Asjo, B., Fenyo, E.M. (1987) Infection of brain-derived cells with the human immunodeficiency virus. J. Virol. 61:1244-1247.

33. Dewhurst, S., Sakai, K., Bresser, J., Stevenson, M., Evinger-Hodges, M.J., Volsky, D.J. (1987) Persis-tent productive infection of human glial cells by human immunodeficiency virus (HIV) J. Virol. 61:3774-3782.

34. Harouse, J.M., Kunsch, C., Hartle, H.T., Laughlin, M.A., Hoxie, J.A., Wigdahl, B., Gonzalez-Scarano, F. (1989) CD4-independent infection of human neural cells by human immunodeficiency virus type 1. J. Virol. 63:2527-2533.

35. Li, X.L., Moudgil, T., Vinters, H.V., Ho, D.D. (1990) CD4-independent, productive infection of a neuronal cell line by human immunodeficiency virus type 1. J. Virol. 64:1383-1387.

36. Nelson, J.A., Wiley, C.A., Reynolds-Kohler, C., Reese, C.E., Margaretten, W., Levy, J.A. (1988) Human immunodeficiency virus detected in bowel epithelium from patients with gastrointestinal symptoms. Lancet i:259-262.

37. Pomerantz, R.J., Kuritzkes, D.R., de la Monte, S.M. et al. (1987) Infection of the retina by human immunodeficiency virus type I. N. Eng. J. Med. 317:1643-1647.

38. Pomerantz, R.J., de la Monte, S.M., Donegan, S.P., et al. (1988) Human immunodeficiency virus (HIV) infection of the uterine cervix. Ann. Intern. Med. 108:321-326.

39. Wiley, C.A., R.D. Schrier, R.D., Nelson, J.A., et al. (1986) Cellular localization of human immunodeficiency virus infection within the brains of acquired immune deficiency syndrome patients. Proc. Natl. Acad. Sci. USA 83:7089-7093.

40. Zucker-Franklin, D., Cao, Y.Z. (1989) The megakaryocytes of human immunodeficiency virus-infected individuals express viral RNA. Proc. Natl. Acad. Sci. USA 86:5595-5599.

41. Tateno, M., Gonzalez-Scarano, F., Levy, J.A. (1989) Human immunodeficiency virus can infect CD4-negative human fibroblastoid cells. Proc. Natl. Acad. Sci. USA 86:4287-4290.

42. Pauza, C.D., J.E. Galindo, J.E., Richman, D.D. (1990) Reinfection results in accumulation of unintegrated viral DNA in cytopathic and persistent human immunodeficiency virus type 1 infection of CEM cells. J. Exp. Med. 172:1035-1042.

43. Dalgleish, A.G., Beverley, P.C.L., Clapham, P.R., et al. (1984) The CD4 (T4) antigen is an essential component of the receptor for the AIDS retrovirus. Nature 312:763-767.

44. Folks, T.M., Powell, D., Lightfoote, M. et al. (1986) Biological and biochemical characterization of cloned leu-3⁻ cell surviving infection with the acquired immune deficiency syndrome retrovirus. J. Exp. Med. 164:280-290.

45. Stevenson, M., Zhang, X., Volsky, D.J. (1987) Downregulation of cell surface molecules during noncytopathic infection of T cells with human immunodeficiency virus. J. Virol. 61:3741-3748.

46. Mullins, J.I., Chen, C.S., Hoover, E.A. (1986) Disease-specific and tissue-specific production of unintegrated feline leukemia virus variant DNA in feline AIDS. Nature 319:333-336.

47. Pang, S., Koyanagi, Y., Miles, S. (1990) High levels

of unintegrated HIV-1 DNA in brain tissue of AIDS dementia patients. Nature 343:85-89.

48. Robinson H.L., Zinkus, D.M. (1990) Accumulation of human immunodeficiency virus type 1 DNA in T cells: result of multiple infection events. J. Virol. 64:4836-4841.

49. Besansky, N.J., Butera, S.T., Sinha, S., Folks, T.M. (1991) Unintegrated human immunodeficiency virus type 1 DNA in chronically infected cell lines is not correlated with surface CD4 expression. J. Virol. 65:2695-2698.

50. Kowalski, M., Potz, J., Basiripour, L., et al. (1987) Functional regions of the envelope glycoprotein of human immunodeficiency virus. Science 237:1351-1355.

51. McCune, J.M., Rabin, L.B., Feinberg, M.B., et al. (1988) Endoproteolytic cleavage of gp160 is required for the activation of human immunodeficiency virus. Cell 53:55-67.

52. Gelderblom, H.R., Reupke, H. Pauli, G. (1985) Loss of envelope antigens of HTLV-III/LAV, a factor in AIDS pathogenesis?. Lancet ii:1016-1017.

53. Gelderblom, H.R., E.H.S. Hausmann, E.H.S., Ozel, M., Pauli, G., Koch, M.A. (1987) Fine structure of human immunodeficiency virus (HIV) and immunolocalization of structural proteins. Virology 156:171-176

54. Lifson, J.D., Hwang, K.M., Nara, P.L., et al. (1988) Synthetic CD4 peptide derivatives that inhibit HIV infection and cytopathicity. Science. 241:712-716.

55. Nara, P.L., Hwang, K.M., Rausch, D.M., Lifson, J.D., Eiden, L.E. (1989) CD4 Antigen-based antireceptor peptides inhibit infectivity of human immunodeficiency virus in titer at multiple stages of the viral life cycle. Proc. Natl. Acad. Sci. USA 86:7139-7143.

56. Byrn, R.A., Mordenti, J., Lucas, C., et al. (1990) Biological properties of CD4 immunoadhesion. Nature 344:667-670.

57. Capon, D.J., Chamow, S.M., Mordenti, J., et al. (1989) Designing CD4 immunoadhesins for AIDS therapy. Nature 337:525-531.

58. Traunecker, A., Schneider, J., Kiefer, H., Karjalainen, K. (1989) Highly effective neutralization of HIV with recombinant CD4-immunoglobulin molecules. Nature 339:68-70.

59. Berman, P.W., Gregory, T.J., Riddle, L., et al. (1990) Protection of chimpanzees from infection by HIV-1 after vaccination with recombinant glycoprotein gp120 but not gp160. Nature. 345:622-625.

60. Goudsmit, J., Debouck, C., Meleon, R.H., et al. (1988) Human immunodeficiency virus type 1 neutralization epitope with conserved architecture elicits early type-specific antibodies in experimentally infected chimpanzees. Proc. Natl. Acad. Sci. USA 85:4478-4482.

61. Javaherian, K., Langlois, A.J., McDanal, C., et al. (1989) Principle neutralizing domain of the human immunodeficiency virus type 1 envelope protein. Proc. Natl. Acad. Sci. USA 86:6768-6772.

62. Javaherian, K., Langlois, A.J., LaRosa, G.J., et al.

(1990) Broadly neutralizing antibodies elicited by the hypervariable neutralizing determinant of HIV-1. Science. 250:1590-1593.

63. Parker, T.J., M.E. Clark, M.E., Langlois, A.J., et al. (1988) Type-specific neutralization of the human immunodeficiency virus with antibodies to env-encode synthetic peptides. Proc. Natl. Acad. Sci. USA 85:1932-1936.

64. Putney, S.D., Matthews, T.J., Robey, W.G., et al. (1986) HTLV-III/LAV-neutralizing antibodies to an E. coli-produced fragment of the virus envelope. Science 234:1392-1395.

65. Rusche, J.R., Javaherian, K., McDanal, C., et al. (1988) Antibodies that inhibit fusion of human immunodeficiency virus-infected cells bind a 24-amino-acid sequence of the viral envelope, gp120. Proc. Natl. Acad. Sci. USA 85:3198-3202.

66. Goudsmit, J., Kuiken, C.L., Nara, P.L., et al. (1989) Linear versus conformational variation of V3 neutralization domains of HIV-1 during experimental and natural infection. AIDS 3(Suppl. 1):S119-S123.

67. Haigwood, N.L., Barker, C.B., Higgins, K.W., et al. (1990) Evidence for neutralizing antibodies directed against conformational epitopes of HIV-1 gp120. Vaccines 90:313-320.

68. Ho, D.D., McKeating, J.A., Ling, X.L., et al. (1991) Conformational epitope on gp120 important in CD4 binding and human immunodeficiency virus type 1 neutralization identified by a human monoclonal antibody. J. Virol. 65:489-493.

69. Nara, P.L., Smit, L., Dunlop, N., et al. Fischinger and J. Goudsmit. (1990) Emergence of viruses resistant to neutralization by V3-specific antibodies in experimental human immunodeficiency virus type 1 IIIB infection of chimpanzees. J. Virol. 64:3779-3791.

70. Chanh, T.C., Dreesman, G.R., Kanda, P. et al. (1986) Induction of anti-HTLV-III/LAV neutralizing antibodies by synthetic peptides. EMBO J. 5:3065-3071.

71. Ho, D.D., Sarngadharan, M.G., Hirsch, M.S., et al., et al. (1987) Human immunodeficiency virus neutralizing antibodies recognize several conserved domains on the envelope glycoproteins. J. Virol. 61:2024-2028.

72. Ho, D.D., Kaplan, J.C., Rackauskas, I.E., Gurney, M.E. (1988) Second conserved domain of gp120 is important for HIV infectivity and antibody neutralization. Science 239:1021-1023.

73. Lifson, J., Coutre, S., Huang, E., Engleman, E. (1986) Role of envelope glycoprotein carbohydrate in human immunodeficiency virus (HIV) infectivity and virus-induced cell fusion. J. Exp. Med. 164:2101-2120.

74. Hansen, J.E., Nielsen, C.M., Nielsen, C., et al. (1989) Correlation between carbohydrate structures on the envelope glycoprotein gp120 of HIV-1 and HIV-2 and syncytium inhibition with lectins. AIDS 3:635-641.

75. Hansen, J.E.S., Clausen, H., Nielsen, C., et al. (1990) Inhibition of human immunodeficiency virus (HIV) infection in vitro by anticarbohydrate monoclonal

antibodies: peripheral glycosylation of HIV envelope glycoptorein gp120 may be a target for virus neutralization. J. Virol. 64:2833-2840.

76. Fenouillet, E., Clerget-Raslain, B., Gluckman, J.C., et al. (1989) Role of N-linked glycans in the interaction between the envelope glycoprotein of human immunodeficiency virus and its CD4 cellular receptor. J. Exp. Med. 169:807-822.

77. Matthews, T.J., Weinhold, K.J., Lyerly, H.K., et al. (1987) Interaction between the human T-cell lymphotropic virus type IIIB envleope glycoptotein gp120 and the surface antigen CD4: role of carbohydrate in binding and cell fusion. Proc. Natl. Sci. USA 84:5424-5428.

78. Alizon, M., Wain-Hobson, S., Montagnier, L., Sonigo, P. (1986) Genetic variability of the AIDS virus: nucleotide sequence analysis of two isolates from African patients. Cell 46:63-74.

79. Zvelebil, M.J.J.M., Sternberg, M.J.E., Cookson, J., Coates, A.R.M. (1988) Predictions of linear T-cell and B-cell epitopes in proteins encoded by HIV-1, HIV-2 and SIV$_{MAC}$ and the conservation of these sites between strains. FEBS Lett. 242:9-21.

80. Kozarsky, K., Penman, M., Basiripour, L., et al. (1989) Glycosylation and processing of the human immunodeficiency virus type 1 envelope protein. J. Acquired Immune Defic. Syndr. 2:163-169.

81. Mizuochi, T., Spellman, M.W., Larkin, M., Solomon, J., Basa, L.J., Feizi, T. (1988) Carbohydrate structures of the human-immunodeficiency-virus (HIV) recombinant envelope glycoprotein gp120 produced in Chinese-hamster ovary cells. Biomed. Chromatogr. 2:260-266.

82. Geyer, H., Holschbach, C., Hunsmann, G., Schneider, J. (1988) Carbohydrates of human immunodeficiency virus. Structures of oligosaccharides linked to the envelope glycoprotein 120. J. Biol. Chem. 263:11760-11767.

83. Gruters, R.A., Neefjes, J.J., Tersmette, M., et al. (1987) Interference with HIV-induced syncytium formation and viral infectivity by inhibitors of trimming glucosidase. Nature 330:74-77.

84. Robinson, E.R., Jr., Montefiori, D.C., Mitchell, W.M. (1987) Evidence that mannosyl residues are involved in human immunodeficiency virus type 1 (HIV-1) pathogenesis. AIDS Res. Hum. Retroviruses 3:265-281.

85. Fenouillet, E., Gluckman, J.C., Bahraoui, E. (1990) Role of N-linked glycans of envelope glycoproteins in infectivity of human immunodeficiency virus type 1. J. Virol. 64:2841-2848.

86. Kumarasamy, R., Blough, H.A. (1985) Galactose-rich glycoproteins are on the cell surface of herpes virus-indected cells. I. Surface labeling and serial lectin-binding studies of Asn-linked oligosaccharides of glycoprotein g C. Arch. Biochem. Biophys. 236:593-602.

87. Ray, E.K., Blough, H.A. (1978) The effect of herpes virus infection and 2-deoxy-D-glucose on glycosphingolipids in BHK-21 cells. Virology 88:118-127.

88. Adachi, M., Hayami, M., Kashiwagi, N., et al. (1988) Expression of LeY antigen in human immunodeficiency virus-infected human T cell lines and in peripheral lymphocytes of patients with acquired immune deficiency syndrome (AIDS) and AIDS-related complex (ARC) J. Exp. Med. 167:323-331.

89. Andrews, P.W., Gonczol, E., Fenderson, B.A., et al. (1989) Human cytomegalovirus induces stage-specific embryonic antigen 1 in differentiating human teratocarcinoma cells and fibroblasts. J. Exp. Med. 169:1347-1359.

90. Montefiori, D.C., Robinson W.E., Mitchell, W.M. (1988) Role of N-glycosylation in pathogenesis of human immunodeficiency virus type 1. Proc. Natl. Acad. Sci. USA 85:9248-9252.

91. Lasky, L.A., Nakamura, G., Smith, D.H., et al. (1987) Delineation of a region of the human immunodeficiency virus type 1 gp120 glycoprotein critical for interaction with the CD4 receptor. Cell 50:975-985.

92. Bahraoui, E., Clerget-Raslain, B., Chapuis, F., et al. (1988) A molecular mechanism of inhibition of HIV-1 binding to CD4+ cells by monoclonal antibodies to gp110. AIDS 2:165-169.

93. Linsley, P.S., Ledbetter, J.A., Kinney-Thomas, E., Hu, S.L. (1988) Effects of anti-gp120 monoclonal antibodies on CD4 receptor binding by the env protein of human immunodeficiency virus type 1. J. Virol. 62:3695-3702.

94. Ruprecht, R.M., Mullaney, S., Andersen, J., Bronson, R. (1989) In vivo analysis of castanospemine, a candidate antiviral agent. J. Acquired Immune Defic. Syndr. 2:149-157.

95. Pincus, S.H., Wehrly, K., Chesebro, B. (1989) Treatment of HIV tissue culture infection with monoclonal antibody-ricin A chain conjugates. J. Immunol. 142:3070-3075.

96. Abe, K., Levery, S.B., Hakomori, S.K. (1984) The antibody specific to type 1 chain blood group A determinant. J. Immunol. 132:1951-1954.

97. Olofsson, S., Blomberg, J., Lycke, E. (1981) O-glycosidic carbohydrate-peptide linkages of herpes simplex virus glycoproteins. Arch. Virol. 70:321-330.

98. Hakomori, S.I. (1985) Aberrant glycosylation in cancer cell membranes as focused on glycolipids: overview and perspectives. Cancer Res. 45:2405-2414.

99. Feizi, T. (1985) Demonstration by monoclonal antibodies that carbohydrate structures of glycoproteins and glycolipids are oncodevelopmental antigens. Nature 314:53-57.

100. Hakomori, S.I., Kannagi, R. (1983) Glycosphingolipids as tumor-associated and differentiation markers. J. Natl. Cancer Inst. 71:231-251.

101. Miyake, M., Zenita, K., Tanaka, O., Okada, Y., Kannagi, R. (1988) Stage-specific expression of SSEA-1-related antigens in the developing lung of human embryos and its relation to the distribution of these antigens in lung cancers. Cancer Res. 48:7150-7158.

102. Solter, D., Knowles, B.B. (1978) Monoclonal anti-

body defining a stage-specific embryonic antigen (SSEA-1) Proc. Natl. Acad. Sci. USA 75:5565-5569.

103. Gooi, H.C., Feizi, T., Kapadia, A., Knowles, B.B., Solter, D., Evans, M.J. (1981) Stage-specific embryonic antigen involves α (1-3) fucosylated type 2 blood group chains.,Nature. 292:156-158.

104. Kannagi, R., Nudelman, E., Levery, S.B., Hakomori, S. (1982) A series of human erythrocyte glycospingolipids reacting to the monoclonal antibody directed to developmentally-regulated antigen, SSEA-1. J. Biol. Chem. 257:14865-14874.

105. Abe, K., J.M. McKibbin and S.I. Hakomori. (1983) The monoclonal antibody directed to difucosylated type 2 chain (Fuca1-2 Galb1-4 [Fuca1-3] GlcNAc; b1-R, Y determinant) J. Biol. Chem. 258:11793-11797.

106. Fukushi, Y., Kannagi, R., Hakomori, S., Shepard, T., Kulander, B.G., Singer, J.W. (1985) Localization and distribution of difucoganglioside (V1³NeuAcV³III³Fuc₂Lc₆) in normal and tumor tissues defined by its monoclonal antibody FH6. Cancer Res. 45:3711-3717.

107. Hakomori, S., Nudelman, E., Levery, S.B., Kannagi, R. (1984) Novel fucolipids accumulating in human adenocarcinoma I. Glycolipids with di- or trifucosylated type 2 chain. J. Biol.Chem. 259:4681-4685.

108. Kannagi, R., Fukushi, Y., Tachikawa, T., Noda, A., Shin, S., Shigeta, K., Hiraiwa, N., Fukuda, Y., Inamoto, T., Hakomori, S., Imura, H. (1986) Quantitative and qualitative characterization of cancer-associated serum glycoprotein antigens expressing fucosyl or sialosyl-fucosyl type 2 chain polylactosamine. Cancer Res. 46:2619-2626.

109. Shi, Z.R., McIntyre, L.J., Knowles, B.B., Solter, D., Kim, Y.S. (1984) Expression of a carbohydrate differentiation antigen, stage-specific embryonic antigen 1, in human colonic adenocarcinoma. Cancer Res. 44:1142-1147.

110. Itzkowitz, S.H., Shi, Z. R., Kim, Y.S. (1986) Heterogenous expression of two oncodevelopmental antigens, CEA and SSEA-1, in colorectal cancer. Histochem. J. 18:155-163.

111. Andrews, P.W., Goodfellow, P.N., Shevinsky, L.H., Bronson, D.L., Knowles, B.B. (1982) Cell-surface antigens of a clonal human embryonal carcinoma cell line: morphological and antigenic differentiation in culture. Int. J. Cancer 29:523-531.

112. Combs, S.G., Marder, R.J., Minna, J.D., Mulshine, J.L., Polovina, M.R., Rosen, S.T. (1984) Immunohistochemical localization of the immunodominant differentiation antigen lacto-N-fucopentaose III in normal adult and fetal tissues. J. Histochem. Cytochem. 32:982-988.

113. Fox, N., Damjanov, I., Martinez-Hernandez, A., Knowles, B.B., Solter, D. (1981) Immunohistochemical localization of early embryonic antigen (SSEA-1) in postimplantation mouse embryos and fetal and adult tissues. Dev. Biol. 83:391-398.

114. Fox, N., Damjanov, I., Knowles, B.B., Solter, D. (1983) Immunohistochemical localization of the mouse stage-specific embryonic antigen I in human tissues and tumors. Cancer Res. 43:669-678.

115. Yamamoto, M., Boyer, A.M., Schwarting, G.A. (1985) Fucose-containing glycolipids are stage- and region-specific antigens in developing embryonic brain of rodents. Proc. Natl. Acad. Sci. USA 82:3045-3049.

116. Karlsson, K.-A. (1989) Animal glycosphingolipids as membrane attachement sites for bacteria. Ann. Rev. Biochem. 58:309-350.

117. Young, J.A.T. (1988) HIV and HLA similarity. Nature 333:215.

118. Vega, M.A., Guigo, R., Smith, T.F. (1990) Autoimmune response in AIDS. Nature 345:26.

119. Gay, D., Maddon, P., Sekaly, R., Talle, M.A., Godfrey, M., Long, E., Goldstein, G., Chess, L., Axel, R., Kappler, J., Marrack, P. (1987) Functional interaction between human T-cell protein CD4 and the major histocompatibility complex HLA-DR antigen. Nature 328:626-629.

120. Diamond, D.C., Sleckman, B.P., Gregory, T., Lasky, L.A. (1988) Inhibition of CD4+ T cell function by the HIV envelope protein, gp120. J. Immunol. 141: 3715-3717.

121. Rosenstein, Y.S., Burakoff, S.J., Herrmann, S.H. (1990) HIV-gp120 can block CD4-class II MHC-mediated adhesion. J. Immunol. 144:526-531.

122. Weinhold, K.J., Lyerly, H.K., Stanley, S.D., Austin, A.A., Matthews, T.J., Bolognesi, D.P. (1989), HIV-1 gp120-mediated immune suppression and lymphocyte destruction in the absence of viral infection. J. Immunol. 142:3091-3097.

123. Sattenau, Q.J., Dalgleish, A.G., Weiss, R.A., Beverley, P.C.L. (1986), Epitopes of the CD4 antigen and HIV infection. Science 234:1120-1123.

124. Clayton, L.K., Sieh, M., Pious, D.A., Reinherz, E.L. (1989) Identification of human CD4 residues affecting class II MHC versus HIV-1 gp120 binding. Nature 339:548-551.

125. Lamarre, D., Ashkenazi, A., Fleury, S., Smith, D.H., Sekaly, R.P., Capon, D.J. (1989) The MHC-binding and gp120-binding functions of CD4 are separable. Science 245:743-746.

126. Golding, H., Robey, F.A., Gates, F.T., Linder, W., Beining, P.R., Hoffman, T., Golding, B. (1988) Identification of homologous regions in human immunodeficiency virus I gp41 and human MHC class II β 1 domain. J. Exp. Med. 167:914-923.

127. Golding, H., Shearer, G.M., Hillman, K., Lucas, P., Manishewitz, J., Zajac, R.A., Clerici, M., Gress, R.E., Boswell, R.N., Golding, B. (1989) Common epitope in human immunodeficiency virus (HIV) 1-gp41 and HLA class II elicits immunosuppressive autoantibodies capable of contributing to immune disfunction in HIV 1-infected individuals. J. Clin. Invest. 83:1430-1435.

128. Glinsky, G.V. (1992) The blood group antigens (BGA)-related glycoepitope: A key structural determinants in immunogenesis and cancer pathogenesis. Critical Reviews in Oncology/Hematology 12:151-166.

129. Springer, T.A. (1990) Adhesion receptors of the immune system. Nature 346:425-434.
130. Springer, T.A., Lasky, L.A. (1991) Sticky sugars for selectins. Nature 349:196-197.
131. Butcher, E.C. (1991) Leukocyte-endothelial cell recognition: three (or more) steps to specificity and diversity. Cell 67:1033-1036.
132. Larsen, E., Palabrica, T., Sajer, S., Gilbert, G.E., Wagner, D.D., Furie, B.C., Furie, B. (1990) PADGEM-dependent adhesion of platelets to monocytes and neutrophils is mediated by a lineage-specific carbohydrate, LNF III (CD15) Cell, 63:467-474.
133. Phillips, M.L., Nudelman, E., Gaeta, F.C.A., Perez, M., Singhal, A.K., Hakomori, S., Paulson, J.C. (1990) ELAM-1 mediates cell adhesion by recognition of a carbohydrate ligand, sialyl-Lex. Science 250:1130-1132.
134. Polley, M.J., Phillips, M.L., Wagner, E.A., Nudelman, E., Singhal, A.K., Hakomori, S., Paulson, J.C. (1991) CD 62 and endothelial cell-leukocyte adhesion molecule 1 (ELAM-1) recognize the same carbohydrate ligand, sialyl-Lewis X. Proc. Natl. Acad. Sci. USA, 88:6224-6228.
135. Berg, E.L., Robinson, M.K., Mansson, O., Butcher, E.C., Magnani, J.L. (1991) A carbohydrate domain common to both sialyl LeA and SialylLex is recognized by the endothelial cell leukocyte adhesion molecule ELAM-1. J. Biol. Chem. 266:14869-14872.
136. Bevilacqua, M.P., Pober, J.S., Mendrick, D.L., Cotran, R.S., Gimbrone, M.A., (1987) Proc. Natl. Acad. Sci. USA 84:9238-9242.
137. Picker, L.J., Kishimoto, T.K., Smith, C.W., Warnock, R.A., Butcher, E.C. (1991) ELAM-1 is an adhesion molecule for skin-homing T cells. Nature 349:796-799.
138. Shimizu, Y., Shaw, S., Graber, N., Copal, T.V., Horgan, K.J., VanSeventer, G.A., Newman, W. (1991) Activation-independent binding of human memory T cells to adhesion molecule ELAM-1. Nature 349:799-802.
139. Mackay, C.R. (1991) Lymphocyte homing. Skin-seeking memory T cells [news; comment]. Nature 349:737-738.
140. Mackay, C.R., Marston, W.L., Dudler, L. (1990) Naive and memory T cells show distinct pathways of lymphocyte recirculation. J. Exp. Med. 171:801-817.
141. Glinsky, G.V. (1992) The blood group antigen-related glycoepitopes: Key structural determinants in immunogenesis and AIDS pathogenesis. Medical Hypothesis, 1992 (in press)

CHAPTER 9

SERUM GLYCOAMINES AS A PROTOTYPE OF THE NEW FAMILY OF CELL ADHESION INHIBITORS

The glycoamine research during the recent 15 years has focused on four main areas: 1) development of an analytical chromatographic method for quantitation of glycoamines in blood serum followed by preliminary characterization of glyco-amines as a new potential human humoral tumor marker; 2) investigation of certain biochemical properties of glycoamines; 3) investigation of the biological action of glycoamines as a humoral factor that are involved in the pathogenesis of malignant growth; 4) structural characterization of glycoamines as a new class of endogenous biopolymers. A fifth focus was inititated recently on the molecular mechanisms of the biogenesis of glycoamines and glycoamine-synthesizing reactions.

ANALYTICAL CHROMATOGRAPHIC METHOD FOR QUANTITATION OF GLYCOAMINES IN BLOOD SERUM AND PRELIMINARY CHARACTERIZATION OF GLYCOAMINES AS A NEW POTENTIAL HUMAN HUMORAL CANCER MARKER

We have developed a micro-method for quantitation of glycoamines in blood serum culminating in a gradient reversed-phase HPLC analysis.[1,2] Quantitative analysis of glycoamines was carried out on 211 oncological patients and 23 healthy donors. The coefficient of variation of the method varied from 6% to 11% with different HPLC systems. The frequency of elevation (2 sigma > mean) of the glycoamines level was 77% from 172 cancer patients, 11% from 35 patients with benign tumors and 0% in donors. The I and II stages of cancer (breast cancer and gynecologic malignancies) revealed an elevation of glycoamine level in 73% from 49 patients.

The analysis of glycoamines includes two basic steps: ultrafiltration and RP-HPLC analysis of the blood serum ultrafiltrate. Detection is by UV at 230 nm. The quantity of blood serum ultrafiltrate needed for one analysis is 5-10 μl. Commercially available ultrafiltration units and membranes are available (with cut off level of 10 kDa from "Amicon", "Bio Rad", Millipore" and "Supelco"). Changing the analytical gradient HPLC procedure to an isocratic mode will permit one to analyze 10 samples per hour. With gradient HPLC 2 to 4 samples per hour can be assayed. Our experimental data show for breast cancer stages I and II an elevation of 56% (18 patients); with lymph node involvement, 100% (3 patients); without, 47% (15 patients), fibromatosis with proliferation, 75% (4 patients); without 0% (4 patients). Early stages of cancer of female genitals, 84% (31 patients). In 86 oncological patients the level of CEA, CA 15-3 and CA 19-9 was also studied.[3] The frequency of elevation in these groups was: glycoamines, (GA), 89%; CEA, 35%; both, 95% (74 patients); GA, 91%; CA 15-3, 58%; both, 98% (45 patients); GA, 89%; CA 19-9, 61%; both, 100% (18 patients); GA + CEA + CA 15-3 + CA 19-9, 99% (86 patients). The elevations are indicated as 2 sigma above the normal mean (in parenthesis is indicated the number of patients).

INVESTIGATION OF CERTAIN BIOCHEMICAL PROPERTIES OF GLYCOAMINES

One of our most important findings in these studies is the characterization of glycoamines as modifiers of peptide-protein binding in blood-serum,[4] as an inhibitor of the $\alpha 2$-macroglobulin mediated restriction of the thrombin-dependent activation of plasma transglutaminase[5,6] and as an inhibitor of nonenzymatic protein glycation.[1,7-10] Structure-function analysis of aminoglycoconjugates implies a structure-function resemblance between glycoamines and linear polyamines.[1,4,8,11,12] Direct biological action of glycoamines may be attributed not only to the presence of free amino groups, but also to the active group of the glycoamine amino acid side groups and the keto groups of carbohydrates. In the last case, the mechanism of action of glycoamines may be similar to the mechanism of action of carbohydrate aldehydes. A reaction of this type may probably determine the covalent incorporation of glycoamines into proteins. Biochemical and structural analysis of aminoglycoconjugates revealed their polyamine-like properties.[1,4,11,12] This can be attributed to the polyamine-like structural organization of free amino groups of the amino acid in aminoglycocojugates and accounts for the high biological activity of glycoamines. In particular, glycoamines bring about modifications in peptide-protein binding in blood plasma and subsequent alterations in the bioactivity of a wide range of protein and peptide bioregulators—neuropeptides, hormones, peptide growth factors in

malignant growth.[1,4] One of the in vitro macromolecular targets for the modification by aminoglycoconjugates is $\alpha 2$ macroglobulin.[1,5,6,8,13] The biochemical synergism of glycoamines and linear polyamines underlies the concept of the monomolecular humorally mediated mechanism of the formation of biological properties of malignant tumors.[1,4,13]

We have observed[5,6] that $\alpha 2$-macroglobulin suppresses in vitro the thrombin-dependent activation of plasma transglutaminase. The inhibiting capacity of $\alpha 2$ macroglobulin is greater for α-thrombin than for γ-thrombin. Glycoamines and polyamines suppress the ability of $\alpha 2$-macroglobulin to inhibit thrombin in the reaction of plasma protransglutaminase activation in a dose-dependent manner. The glycoamines and polyamine-dependent suppression of the inhibiting capacity of $\alpha 2$-macroglobulin may play a significant role in disturbances of the adhesion properties of cancer cells.

STRUCTURAL CHARACTERIZATION OF GLYCOAMINES AS A NEW CLASS OF ENDOGENOUS BIOPOLYMERS

The chemical structure of glycoamines represents carbohydrate-amino acid conjugates that contain from 5-29 amino acids and from 1-17 carbohydrate residues. Amino acids in glycoamines are joined to the carbohydrate core part of the molecule via ester bonds. The structure of over 53 aminoglycoconjugates isolated from the blood plasma and serum of rodents and humans has been investigated.[1,10-12,14-16]

We can now present two structural models for aminoglycoconjugates purified from blood plasma of oncology patients.[1,10,13,16] These models are based on data of the amino acid and carbohydrate composition, protein sequence, infrared spectroscopy, structural investigation of the carbohydrate core of the molecule, and intermediate products obtained after mild acidic and alkaline hydrolyses. In accordance with the first model, the amino acids are joined to the core carbohydrate components via ester bonds and separate amino acid-carbohydrate blocks are joined in the molecule through the glucose aldehyde by way of the formation of Schiff bases and

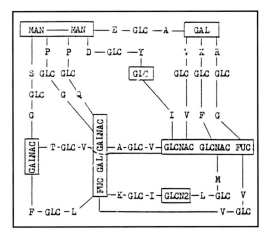

Fig. 1. Structure of the type-1 glycoamine.[13]

Amadori products. (Fig. 1) The structure of the second type of glycoamine differs to a certain extent from this model. According to structural analysis, the carbohydrate blocks are joined via the amino acid "bridges" with the formation of ester bonds, Schiff bases and Amadori products. The monosaccharide components form the peripheral part of the molecule and their reducing ends are available for interaction with 7-amino-4-methylcoumarin (AMC) in the reaction of reducing amination. The reducing ends of disaccharides and trisaccharides are blocked and become available for the interaction with AMC in the reaction of reducing amination after lithium borohydride hydrolysis, which probably destroys bonds that are more stable than Schiff bases, the Amadori product type bonds. (Fig. 2)

Thus, the structural analysis of glycoamines reveals that the amino acids and carbohydrates are linked by labile bonds. At least some of these bonds (Schiff bases and Amadori products) may be formed nonenzymatically. Further investigations showed that glycoamines are able to incorporate monomeric carbohydrates and amino acids by a nonenzymatic mechanism.[9,10] In the case of amino acids, the reaction of nonenzymatic incorporation involves the formation of Schiff bases, and the subsequent spontaneous hydrolysis of Schiff bases accompanied by intramolecular rearrangement, with the formation of an ester bond between the amino acid carboxyl and carbohydrate hydroxyl and "secondary" blocking of the amino acid α-amino group as a result of the interaction with the aldehyde or keto group.

The results of investigations on the structure and biogenesis of glycoamines reveal a unique capacity of these compounds for nonenzymatic modification of the peripheral part of the molecule by way of incorporation of new amino acid and carbohydrate units. The analysis of the structural-function interrelations accomplished using chemical, biochemical, spectroscopic, and biological methods indicate that amino acids and carbohydrates are the main structural-functional determinants of glycoamines. Thus, the structural studies on natural glycoamines and their synthetic analogs, along with the identification of the structural-functional determinants of aminoglycoconjugates, provide the chemical

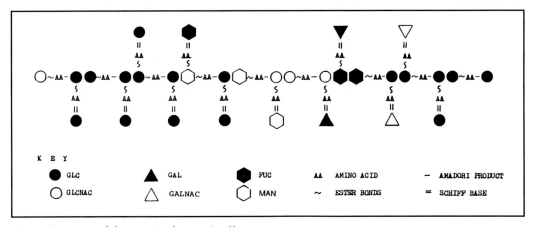

Fig. 2. Structure of the type-2 glycoamine.[13]

basis for the hypothesis of humoral molecular "imprints."[1,9,10,13,17-20] Investigation of the biological properties of glycoamines reveal the carbohydrate component as the key structural-functional determinant. One of the main reactions of the biogenesis of glycoamines is the nonenzymatic interaction between the carbohydrate aldehydes and α-amino groups of amino acids.[1,9,10]

INVESTIGATION OF THE BIOLOGICAL ACTION OF THE GLYCOAMINES AS A HUMORAL FACTOR OF CANCER PATHOGENESIS

In this line of experiments, we have focused on the investigation of the possible role of glycoamines as a factor in metastasis. We suggest that glycoamines may be involved in this process through a modification of aggregation and cell association properties of tumor cells.

One of the characteristic features of the growth of malignant tumors cultured in vitro is their ability to form three-dimensional spheroid-like structures the multicell spheroids.[21,22] According to biochemical, immunocytochemical, morphological and cytological criteria, similar spheroid-like formations were observed in rodent and human malignant tumors in vivo. Therefore, the multicell spheroids are considered a cytofunctional analog of the avascular stage of in vitro tumor development.[22] Under our experimental conditions, about 90% of the spheroids formed by tumor cells are in suspension and about 10% are attached to the substrate. The protein fraction from blood serum of tumor-bearers with a molecular weight over 100 kDa stimulates spheroid formation, increasing the number of spheroids in suspension two-fold. (Fig. 3) Glycoamines of less than 10 kDa inhibit tumor spheroid formation by more than 95%, bringing about dramatic changes in adhesion properties of tumor cells and a disaggregation of multicell spheroids. What is the mechanism of this phenomenon? One possible mechanism is the one we have presented in detail previously.[1,13] At that time we presented data on the biochemical synergism of glycoamines and linear polyamines as potential anti-adhesion factors that favor the dissemination of tumor cells.[1,13] Also, primary amine-dependent restraint on the activity of protease inhibitors may play an important role in this effect. Further, in vivo and in vitro investigations on glycoamines have revealed suppression of immunological response to tumor antigens. Thus, it is possible that glycoamines may be a humoral factor contributing to tumor-associated immunosuppression. The stimulation of spheroid formation in suspensions induced by serum proved to be caused by immunoglobulins and is probably associated with reactions of the antigen-antibody type. The glycoamine fraction in blood serum with a molecular weight below 10 kDa inhibits the in vitro reassociation of lymphocyte-macrophage aggregates isolated from the spleen of tumor-bearing animals. These data, in conjunction with the analysis of the glycoamine structure, enabled us to formulate the hypothesis of a monovalent mechanism of antiadhesive and immunosuppressive actions attributed to aminoglycoconjugates of < 10 kDa.[13,17-20]

Of great importance is the phenomenon of complete correspondence between the effect of blood-serum fractions on spheroid formation in vitro and their effect on metastatic spread in vivo after intravenous inoculation of tumor cells. The fraction of blood serum with a molecular weight over 100 kDa increases the number of tumor-cell colonies in the lungs over two-fold, whereas the fractions with a molecular weight below 10 kDa and HPLC purified glycoamines considerably inhibit the formation of lung metastases and remove metastasis stimulation by the high molecular weight fraction of blood serum. (Fig. 3) Amino acids and carbohydrates have been identified as the main structural-functional determinants of the glycoamines. The functional identification of carbohydrate determinants was performed using the hemagglutination inhibition test, the peptide- protein binding assay, lectin chromatography, and in a test of the inhibition of aggregate formation by tumor cells.[8] The inhibition of tumor cell aggregation in vitro by glycoamines could be prevented by influence of temperature and/or various types of hydrolysis, may be mediated by synthetic fructosyl amino acids or glycoesters, (Fig 4)

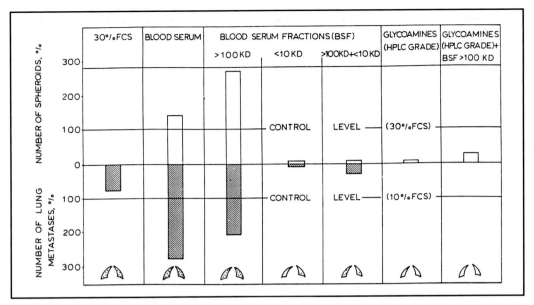

Fig. 3. Effects of glycoamines and fractions of blood serum on the spheroid formation in vitro and metastatic spreading in vivo.[13,20]

and involved galactose as a terminal carbohydrate residue.[20] These experiments have been done in different experimental models of malignant growth.

We have investigated the effect of combined administration of glycoamines and spermidine on the metastatic spread of Lewis carcinoma under the ablation of primary tumor. Administration of glycoamines and spermidine stimulated lung metastases threefold in mice. We have shown that the effect of stimulation of metastatic spread is exhibited under the condition that the administration of glycoamines and spermidine starts in the presence of primary tumor. The effect disappears when the administration starts one day after the ablation of primary tumor. Thus, the glycoamine and spermidine-dependent stimulation of metastatic spreading is associated with the activation of tumor cell dissemination from the primary focus.[1,13, 20]

Two important conclusions can be drawn from the experiments on the effect of glycoamines and fractions of blood serum on cell aggregation in vitro and on metastasis. First, the formation of oncoaggregates and re-association of immunoaggregates occurs with the involvement of common or similar structural determinants, probably of carbohydrate nature. Second, the humoral (serum) constituents, glycoamines and high molecular weight proteins of blood serum that are able to recognize these structural determinants are present not only in the cancer serum, but also to a lesser extent in normal blood serum. Such structural determinants participating in the homotypic aggregation of histogenetically different types of cells may be, in particular, the carbohydrate determinants of the blood-group antigen (BGA) related glycoantigens, since normal blood serum contains anticarbohydrate antibodies (including those against BGA-related glycoepitopes), and preliminary data indicate the involvement of BGA-related glycoepitopes in the inhibition of hemagglutination by glycoamines.

At present, most of the clinically used humoral tumor markers are glycomacromolecules (CEA, mucins, CA 19-9, CA15-3, CA 26); many of them contain the BGA-related glycoepitopes (Lewis family, T, Tn, I)[23-28] and reveal immunosuppressive properties.[29] Blood serum from most healthy individuals contain natural anti-carbohydrate antibodies, including those against the BGA-related glycoepitopes (anti-A, B, H, O, I, and Lewis family; anti-Tn-antibodies; antiasialo GM_2 ([GalNac β (1-4) Galβ (1-4) Glc-Cer]).[30]

Blood serum from oncological patients reveals a decrease in the titers of anti-T, anti-Tn, anti-I and anti-Forssman anticarbohydrate antibodies.[23,31,32] Finally, the most characteristic manifestation of aberrant glycosylation of cancer cells is neosynthesis (or ectopic synthesis), the synthesis of incompatible antigens and incomplete synthesis (with or without the accumulation of precursors) of the BGA-related glycoepitopes.[33-35]

BGA-related glycoepitopes are directly involved in the homotypic (tumor cells, embryonal cells) and heterotypic (tumor cells-normal cells) formation of cellular aggregates, e.g., Lewis X antigens; H-antigens, polylactosamine sequences; and T-and Tn- antigens, which was demonstrated in different experimental systems.[31,36-39] BGA-related alterations in the tissue glycosylation pattern are detected in benign (premalignant) tumors with high risk of malignant transformation, in primary malignant tumors, and in metastases,[35,40-45] i.e., they were demonstrated as typical alterations in different stages of tumor progression. Lymphoid cells (lymphocytes, granulocytes, monocytes) are Lewis family glycoepitope-positive and contain T-antigen[46-48] (demasked after treatment with neuraminidase). Based on these facts, we have suggested that the processes of thymic education and antigen presentation are accompanied by certain changes in the glycosylation patterns of the cellular membranes. These changes include the presentation of the BGA-related glycoepitopes—in particular, of their cryptic forms.[13,17-20]

Thus, glycoamines are a newly recognized class of endogenous small molecular weight biopolymers consisting of amino acids and sugar units with covalent structure of ester, Schiff bases and Amadori bonding. They have been investigated during the last 15 years as physiological components of human and rodent blood serum that merit interest as potential human humoral tumor makers.[1-3,7,14,49] Structurally the glycoamines represent carbohydrate-amino acid conjugates containing from 5-29 amino acids and from 1-17 carbohydrate residues.[1,11-15] The level of these substances is substantially increased in blood serum from humans and animals with different forms of malignant solid tumors and leukemias. The chemical structure of glycoamines reveal mono-, di- and trisaccharides bound by ester links to the amino acids and assembled into higher molecular weight compounds via the formation of Schiff base and Amadori product-type bonds with the involvement of

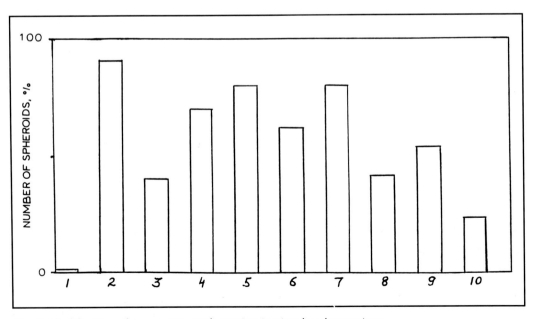

Fig. 4. Modification of oncoaggregate formation in vitro by glycoamines.

the amino groups of amino acids and alde-hyde or keto groups of the carbohydrates.[1,11-16] Elevated level of glycoamines in the body fluids of cancer patients may provide one of the important factors that determine the malignant behavior of tumors in vivo.

The Concept of the New Faminly of Antimetastatic Drugs

The most important presently known biological function of glycoamines is their activity as an in vitro inhibitor of cell aggre-gation and in vivo antimetastatic agent in an experimental model of metastasis. Those ob-servations are particularly important since current concepts have developed regarding the great potential of synthetic and natural carbohydrate cell adhesion inhibitors as antimetastatic, anti-AIDS and antiinflammatory durgs. In our studies we have used the methylcholantrene (MX)-induced fibrosar-coma in BAL B/C mice as a source of tumor cells for the multi-cell aggregation assay. After primary induction of these tumors by MX, up to 25 subsequent syngenic transplanta-tions did not produce any significant changes in in vitro cell aggregation properties. These tumors express the sialosyl -LeX antigen which is well known as a human tumor-associated carbohydrate antigen that is involved in homotypic (cancer cell-cancer cell) and het-erotypic (cancer cell-normal cell), e.g., plate-lets, leukocytes, cell aggregation. The Gal residues have been identified as terminal car-bohydrate determinants involved in cell ag-gregation of this experimental tumor that is identical to many human cancers; this experi-mental tumor shows a broad sensitivity to cell aggregation inhibitory activity of glyco-amines regardless of the source of compound purification, e.g., human or rodent serum, and including synthetic structural analogs of glycoamines. We have finalized this study for the above described syngenic animal model by purification and structural characteriza-tion of individual glycoamines in mice serum that are responsible for in vitro tumor cell aggregation inhibitory activity and confirmed this observation by bioassay of corresponding synthetic structural analogs of glycoamines.[50]

The objective of current research is to do the same in relation to human cancer.

We do not have information regarding the specificity of this phenomenon for indi-vidual human tumors. However, the many similarities between human cancer and our experimental model of in vitro cell aggrega-tion, as well as our successful experience in an experimental model of cancer, provide the opportunity for expansion of this study with regard to human cancer, particularly for analy-sis of this phenomenon in each individual system of tumor and serum specimen. We also plan to replace the current model of in vitro tumor cell aggregation with well-char-acterized tumor cell lines that express Gal-binding lectins and in which the Gal residues have been shown to be involved in cell aggre-gation. We have recently identified in the glycoamine fraction of cancer serum a num-ber of compounds that are elevated and/or present exclusively in cancer serum. (Fig. 5) It has been observed that the compounds (Figs. 6 & 7) are extremely sensitive to 10% NH_4OH hydrolysis at 41°C during 16h. Be-cause similar treatment dramatically reduced the activity of low molecular weight cancer cell adhesion inhibitor(s) in serum[20,50] it is possible that this compound is involved in in vitro inhibition of cancer cell aggregation.

In summary, different molecular isoforms of glycomacromolecules in serum contain cell-originated glycodeterminants which are in-volved in cell recognition, association and aggregation. These glycomacromolecules and serum carbohydrate-binding proteins, e.g., lectins, anticarbohydrate antibodies, etc., play an opposite roles as inhibitors (low molecular weight glycoamines) and/or stimulators (high molecular weight multiglycoepitope-contain-ing serum macromolecules and naturally oc-curring anticarbohydrate antibodies) of cell aggregation. Human cancer is accompanied by changes in the quantitative and qualita-tive characterstics of these serum macromol-ecules, generally in the direction of excessive accumulation of the activities that support tumor cell aggregation in blood vessels and metastasis. These conditions specifically are being changed by surgical elimination of pri-mary tumor because as we have shown[1,49] in

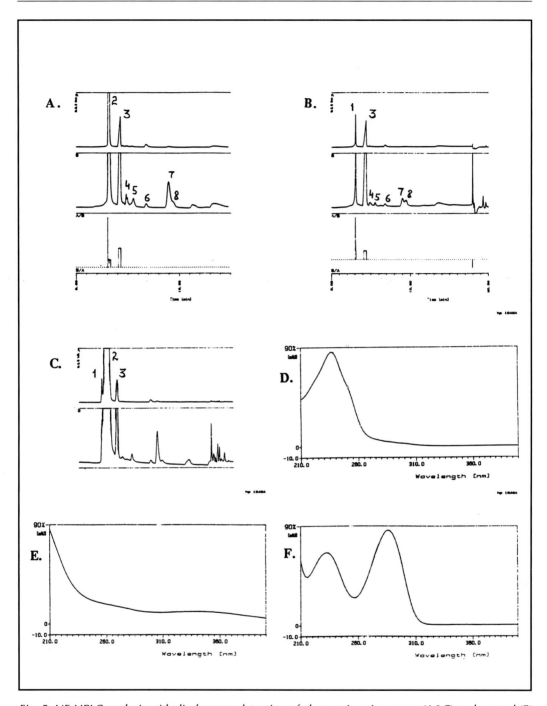

Fig. 5. UF-HPLC analysis with diode-array detection of glycoamines in cancer (A&C) and control (B) blood serum samples. Diode-array spectra of compounds 1, 2, & 3 (panel C) are presented on figures E, D & F, correspondingly. Note that compound(s) with spectral characteristics presented in panel D has not been detected in the control serum.

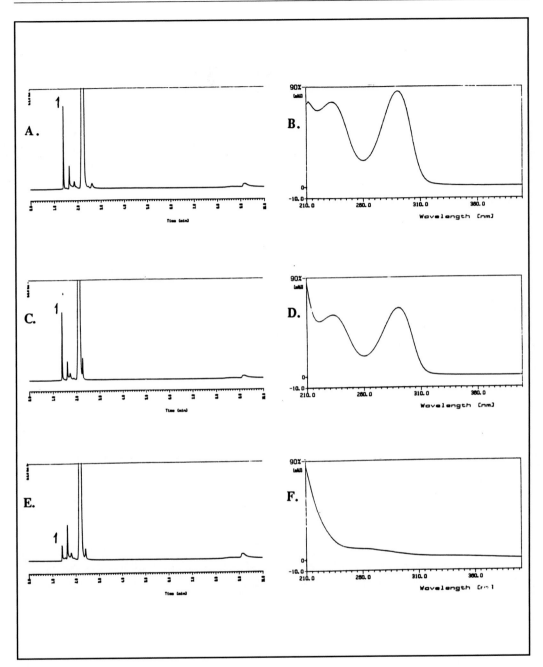

Fig. 6. HPLC analysis with diode-array detection of the glycoamine fraction of cancer serum after hydrolysis in 10% NH$_4$OH at 41°C at 3 minutes (panel A), 20 minutes (panel C) and 16 hours (panel E), and correspondingly diode-array spectra of compound(s) 1 (B, D & F).

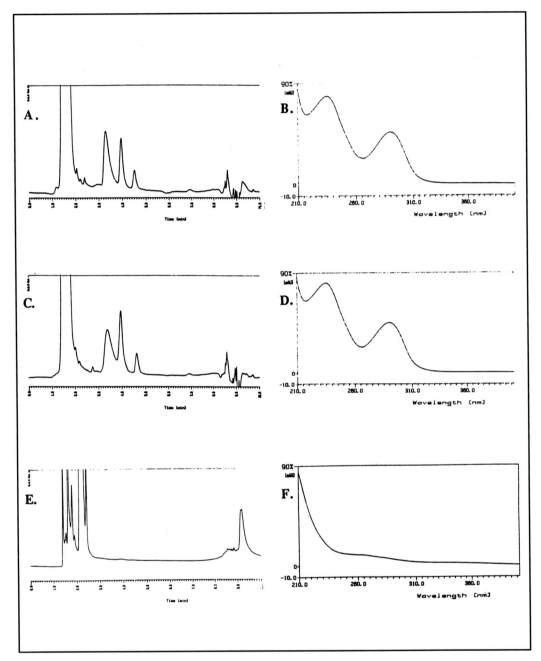

Fig. 7. HPLC analysis with diode-array detection of the glycoamine fraction of cancer serum after storage at -20°C (panel A) incubation at 41°C at 16 hours (panel C), and hydrolysis in 10% NH₄OH at 41°C at 16 hours (panel E), and correspondingly diode-array spectra (B, D & F) of compound(s) of interest.

the early period after surgery (2-3 days) that the level of glycoamines (the major cell aggregation and metastasis inhibitory component in serum) decreases dramatically in the serum of oncological patients. The changes in the levels of stimulatory macromolecules are supposedly less radical since some of them (anticarbohydrate antibodies) are present premanently in serum of all healthy individuals[30,31] and it is well known that high molecular weight proteins in serum have a relatively long half-life time.

Initial observations have shown the effectiveness of these types of synthetic and natural molecules with biospecificity for glycodeterminants of cancer cell adhesion as inhibitors of in vitro cell aggregation and potential therapeutic agents in vivo.[1,8,9,13,17-20,36,37,51-60] It has been shown that specific carbohydrate units of glycosphingolipids as well as glycosphingolipid-containing liposomes significantly suppressed B16 melanoma lung metastasis.[60] Effective inhibition of an experimental metastasis has been shown by peptide cell adhesion determinants[61-63] as well as by monoclonal antibody against lectins that participate in tumor cell aggregation (see above). However, the disadvantages of a secondary immunological reaction and the nonspecific nature of these molecules for tumor cells have blocked their potential therapeutic use.

Cell aggregation glycodeterminants never has been a target for antimetastatic drug development. But at least some of them have a unique design for anticancer drug potential: they have been detected in from 75% to more that 90% of human carcinomas and never been detected in normal healthy tissues.[31,48] The use of monoclonal antibodies and/or active immunization against these glycodeterminants obviously has very limited potential, since, as we already have mentioned, serum of healthy individuals contain naturally occurring anticarbohydrate antibodies of the same specificity[30,31] and immunoglobulins, particularly IgM, do not readily permeate blood vessels and have no access to extravascular tumor cells. It has been considered that naturally occurring anticarbohydrate antibodies may play a key role in cancer pathogenesis as a major immunoselective factor of metastasizing cancer cells during tumor progression,[13,20] and it has been shown that mAb against glycodeterminants of cancer cell aggregation did induce a strong in vitro aggregation of human tumor cells.[64]

However, many key questions remain unanswered. How specific are these phemonema for individual human tumors? What is the relationship if any, between expression of glycodeterminants of cancer cell aggregation on the cell surface of tumor and their presence in extracellular medium, particularly in serum? What is the chemical nature of low molecular weight cancer cell aggregation inhibitor(s) that present in human serum? Is it similar (or identical) to rodent cancer cell aggregation inhibitor which presents in rodent serum and has proven to be glycoamine? How will surgery, chemotherapy, and hormonal treatment change the activity and/or quantity of serum factor(s) that modify cancer cell aggregation? What is the diagnostic and prognostic significance, if any, of those parameters? We will address many of these problems in our ongoing research.

We will initiate the experimental verification of the hypothesis that there is a key regulatory element in the development control program which functions as a cell-dependent and cell type-specific "education mechanism" of extracellular soluble macromolecules regarding both quantity and biospecificity of glycodeterminants of cell-cell recognition and adhesion. Since we have recently shown that certain synthetic structural analogs of glycoamines with previously undescribed chemical structure act as inhibitors of in vitro cell aggregation, this project will lead to the discovery of new cellular receptor system(s), and/or new molecular regulatory mechanisms of cell-cell type specific recognition.

For verification of the "new cellular receptor system" hypothesis, a selected number of synthetic compounds with high biological activity will be designed as radiolabeled substances and radio-receptor analysis (RRA) with corresponding target cells, tissues, and/or purified membrane fraction will be performed. Since some of the synthetic structural analogs of glycoamines with previously undescribed structures, e.g., glycoesters, fructosyl amino acids) function as in vitro inhibitors of cell aggregation, those compounds will be primary candidates for RRA.

The results of investigations on the structure and biogenesis of glycoamines reveal a unique capacity of these compounds for nonenzymatic modification of the peripheral part of the molecule by way of incorporation of new amino acid and carbohydrate units. On the other hand, the high valence of glycoepitopes is critically important for the biological action of oligosaccharides;[65] and the multivalent clustered structure of glycodeterminants is important for immune system recognition and these structural arrangements seem to be important for recognition of carbohydrates by cells or binding proteins.[66-68] For example, the low affinity of anticarbohydrate antibody is greatly enhanced if the antigenic determinants are bi- or multivalent.[46] Binding of antigen by two or multiple combining sites is favored over binding at a single site by a factor of 10^3 - 10^4.[69] As a result of glycosidase cleavage specific oligosaccharides will be released from cell membrane glycomacromolecules and subsequently they could interact nonenzymatically with extracellular glycoamines. Thus, multivalent, clustered and highly active and biospecific glycopolymers could be formed. These molecules could be effector molecules directly responsible for glycoamine-dependent inhibition of cell aggregation.

The above considerations provide a basis for our previously suggested humoral molecular cell-biospecific imprint hypothesis[13,20] that described the "new molecular regulatory mechanism" of cell-cell recognition and aggregation. Experimental verification of this hypothesis will be performed using in vitro cell aggregation assays, synthetic structural analogs of glycoamines, specific oligosaccharides that are expressed on the cell surface of particular cell lines and are responsible for aggregation of those cells, and specific synthetic compounds that will be designed as glycoamine-oligosaccharide copolymers which contain corresponding glycoamines as the core part of the effector molecule, and oligosac-

charides as the peripheral bioaddressing part. The substitution of polylysine for the glycoamine core component in synthetic molecules will provide some additional possibilities for study of structure-function relations.

In summary, histogenetically different normal (embryonal, lymphoid) and a wide range of cancer cells and tissues express either transitory (normal) or permanently (cancer) the definite BGA-related glycodeterminants that are involved in initial early stages of a multistep "cascade" cell adhesion mechanism. Like embryogenesis, tissue repair, regeneration and remodeling could be accompanied by expression of BGA-related glycoepitopes at definite stages of cell differentiation when the "sorting-out" behavior of one homotypic cell population from heterotypic assemblage of cells occurs. However, many key questions remain unanswered. How specific are these phemonema for individual tissues and cell lines? What are the relationships if any, between expression of glycodeterminants of cell aggregation on the cell surface and their presence in extracellular medium, particularly in serum? How will development (embryogenesis, morpho- and organogenesis) and physiological processes, e.g., immune response, change the activity and/or quantity of serum factor(s) that modify cell aggregation? What is the functional significance, if any, of those changes? We will address many of these problems in our research.

The family of antimetastatic drugs will be designed with biospecificity for cellular glycodeterminants that are involved in homotypic (tumor cell-tumor cell) and heterotypic (tumor cell-normal cell) cell-cell recognition, association and aggregation. These drugs will be composed of several synthetic, low molecular weight compounds which contain different core structures and appropriate glycoepitopes as peripheral structures of effector molecules.

CORE COMPONENT OF SYNTHETIC COMPOUNDS

As a core structure for synthetic compounds we will utilize di- and polylysine and amino acid-glucose copolymers. We will use these types of carrier molecules for development of multivalent and clustering structures, since the multivalent clustered structure of glycoepitopes is important for immune system recognition (particularly T-, Tn- and sialosyl-Tn- glycoantigens) and these structural arrangements seem to be important for recognition of carbohydrates by cells or binding proteins.[66-68] As we already have mentioned, the low affinity of anticarbohydrate antibodies is greatly enhanced if the antigenic determinants are bi- or multivalent.[46] Binding of antigen by two or multiple combining sites is favored over binding at a single site by a factor of 10^3 - 10^4.[69]

L - Lysyllysine is suitable for a backbone which constructs multivalent structures,[57-59] since its three amino groups (one α- and two Σ-amino groups) can be utilized to attach three glycodeterminant clusters and one carboxyl group is available for direct coupling to carrier molecules. Obviously, polylysine molecules with different molecular weights provide additional structural flexibility.

Recently, the chemically defined approach known as multiple antigen peptide system (MAP) for vaccine engineering has been designed to incorporate peptide antigens that give a macromolecular structure without a large protein carrier.[70-72] The core of the MAP, composed of multiple lysines, is a branching backbone. A number of antigen chains—2,4,8,16 or 2^n—can be attached to the lysine tiers in the backbone. The final synthetic molecule, after deprotection and cleavage, may be used as is, thus bypassing the need for conjugation to a large carrier protein such as keyhole limpet hemocyanin (KLH), or bovine serum albumin (BSA). In contrast to the conventional antigen carrier protein conformation where the mass of the carrier dominates the antigen molecule mass, the antigen content of the MAP is typically 80-90% of the molecule mass. It is important that antisera to MAP antigens recognize the native proteins, but do not contain antibodies to the polylysine core.[70] We will utilize the MAP system methodology for development of a multiple glycoepitope system (MGS) for design of branching homo- and heterogamous glycodeterminants containing synthetic molecules with biospecificity for cell-cell recognition carbohydrate structures.

Synthetic structural analogs of the basic structural units of glycoamines will be used as core carrier molecules as well as a peripheral bioactive structures, since glycoamines and their synthetic analogs show strong inhibition of in vitro tumor cell aggregation and metastasis in vivo (see above). We have indicated in our structural studies of the natural glycoamines purified from serum of oncological patients[1,10-12,15,16] that the most representative carbohydrate component of glycoamines is glucose. The amino acids have been identified in purified glycoamines with the following frequency: D (90%); E (90%); S (70%); G (100%); T (60%); A (100%); Y (40%); M (20%); V (70%); P (80%); K (80%); F (40%); I (40%); L (40%); and R (40%).

Correspondingly, synthetic structural analogs of basic structural units of glycoamines (glycoesters, glycosides, Schiff bases, Amadori products, fructosyl amino acids) will comprise amino acid-glucose compolymers with structural characteristics (amino acid: glycose ratio and type of covalent bonding) related to those in natural compounds.[1,10,15,16]

The peripheral bioactive part of the synthetic compounds will be composed according to the specific background information available from the human breast and lung cancer studies (see above).

REFERENCES

1. Glinsky, G. V. (1989) Glycoamines: Biochemistry of a new class of humoral tumor markers. J. Tumor Marker Oncology 4:193-221.
2. Glinsky, G.V., Glinsky, V.V., Surmilo, N.I., Shvachko, L.P., Radavsky, Yu.L., Kukhar, V.P. (1986) Proc. Acad. Sci. USSR. V.228, N 2, p. 495-499, (Rus)
3. Glinsky, G.V., Birkmayer, G.D., Vinnitskaya, A.B., Koritskaya, L.N., Osadchaya, L.P., Ivanova, A.B. (1990) J. Tumor Marker Oncology 5:249.
4. Glinsky, G.V. (1987) J. Tumor Marker Oncology 1:249-294.
5. Glinsky, G.V., Surgova, T.M., Sidorenko, M.V., Vinnitsky, V.B. (1990) J. Tumor Marker Oncology 5:161-66.
6. Surgova, T.M., Glinsky, G.V., Sidorenko, M.D., Shvachko, L.P., Kibirev, V.K., Kukhar, V.P., Vinnitsky, V.B. (1988) Proc. Acad. Sci. Ukr. SSR., B. series, N 5, p. 79-81, (Rus)
7. Glinsky, G.V., Surmilo, N.I., Liventsov, V.V., Sidorenko, M.V. (1990) J. Tumor Marker Oncology

5:250.

8. Glinsky, G. V., Linetsky, M.D., Ivanova, A.B., Osadchaya, L.P., Semyonova-Kobzar, R. A., Surmilo, N. I. (1990) Chemical, biochemical, spectroscopic and biological indentification of structural-functional determinants of glycoamines (aminoglycoconjugates) J. Tumor Marker Oncology 5:250.

9. Glinsky, G.V., Ogordniychuk, A.S., Shilin, V.V., Linetsky, M.D., Liventsov, V.V., Sidorenko, M.V., Surmilo, N.I., Kuchar, V.P (1990) Structural analysis of synthetic structural analogs of glycoamines: contribution into elucidation of the structure and biogenesis of natural aminoglycoconjugates. J. Tumor Marker Oncology 5:250.

10. Glinsky, G.V., Linetsky, M.D., Liventsov, V.V., Surmilo, N.I., Ivanova, A.B., Sidorenko, M.V., Vinnitsky, V.B. (1990) J. Tumor Marker Oncology 5:119-136.

11. Glinsky, G.V. (1988) Proc. Acad. Sci. Ukr. SSR., B series. N 3, p. 69-72, (Rus)

12. Glinsky, G.V. (1988) Proc. Acad. Sci. Ukr. SSR., B series, N 4, p. 43-49.

13. Glinsky, G.V. (1992) Glycoamines: Structural-Functional characterization of a new class of human tumor markers., In: Serological Cancer Markers. Editor: S. Sell. The Humana Press., Totowa, NJ, Chapter 11, p. 233-260.

14. Glinsky, G.V., Vinnitsky, V.B. (1987) Experimental Oncology V.9, N 5, p. 78-80. (Rus)

15. Glinsky, G.V., Linetsky, M.D., Shemyakin, V.V., Markin, V.V., Liventsov, V.V., Ivanova, A.B. (1990) J. Tumor Marker Oncology 5:107-18.

16. Glinsky, G.V., Linetsky, M.D. (1990) J. Tumor Marker Oncology 5:137-60.

17. Glinsky, G.V. (1990) Immunoselective hypothesis of tumor progression. Role aberrant glycosylation, anti-carbohydrate antibodies, extracellular glycomacromolecules and glycoamines. J. Tumor Marker Oncology 5:206.

18. Glinsky, G.V. (1990) Glycoamines, aberrant glycosylation and cancer: a new approach to the understanding of molecular mechanism of malignancy., In: Molecular Oncology. Oncodevelopment proteins and clinical applications. XVIIIth meeting of the International Society for Oncodevelopmental Biology and Medicine. Abstract Book, Moscow, USSR, September 23-27, 1990, p.7.

19. Glinsky, G.V., Semyonova-Kobzar, R.A., Berezhnaya, N.M. (1990) Modification of cellular adhesion, metastasizing and immune response by glycoamines: implication in the pathogenetical role and potential therapeutic application in tumoral disease. J. Tumor Marker Oncology 5:231.

20. Glinsky, G.V. (1992) The blood group antigens (BGA)-related glycoepitopes. A key structural determinants in immunogenesis and cancer pathogenesis. Critical Reviews in Oncology/Hematology 12:151-166.

21. Mueller-Kleiser, W. (1987) Multicellular spheroids. A review on cellular aggregates in cancer research. J.

Cancer Res. Clin. Oncol. 113, 101-122.

22. Sutherland, R.M. (1988) Cell and environment interactions in tumor microregions: the multicell spheroid model. Science 240: 177-184.

23. Dube, V.E. (1987) The structural relationship of blood group-related oligosaccharides in human carcinoma to biological function: A perspective. Cancer Metastasis Rev. 6, 541-557.

24. Kannagi, R., Hakomori, S.I., Imura, H. (1988) In: Altered glycosilation in tumor cells. Editors: Ch. L. Reading, S.I. Hakomori, D.M. Marcus., New York: Alan R. Liss, 279-94.

25. Linsley, P.S., Brown, J.P., Magnani, J.L., Horn, D. (1988) Monoclonal antibodies reactive with mucin glycoproteins found in sera from breast cancer patients. Cancer Res. 48: 2138-48.

26. Magnani, J., Nilsson, B., Brockhaus, M., Zop, J.D., Steplewski, Z., Koprowski, H., Ginsburg, V. (1982) A monoclonal antibody-defined antigen associated with gastrointestinal cancer is a ganglioside containing sialylated lacto-N-fucopentaose II. J.Biol. Chem. 257: 14365-14369.

27. Magnani, J., Steplewski, Z., Koprowski, H., Ginsburg, V. (1983) Identification of the gastrointestinal and pancreatic cancer-associated antigen detected by monoclonal antibody 19-9 in the sera of patients as a mucin. Cancer Res. 43: 5489-5492.

28. Yamashita, K., Totani, K., Kuroki, M., Matsuoka, Y., Ueda, I., Kobata, A. (1987) Cancer Res. 47: 3451-3459.

29. Hakim, A.A. (1984) Neoplasma 31: 385-397.

30. Lloyd, K. O., Old, L.J. (1989) Human monoclonal antibodies to glycolipids and other carbohydrate antigens: dissection of the humoral immune response in cancer patients., Cancer Res. 49:3445-3451.

31. Springer, G.F. (1984) T and Tn, general carcinoma autoantigens. Science 224: 1198-1206.

32. Young, W.W., Hakomori, S.I., Levine, P. (1979) Characterization of anti-Forssman (anti-Fs) antibodies in human sera: their specificity and possible changes in patients with cancer. J. Immunol. 123: 92-96.

33. Hakomori, S.I. (1985) Aberrant glycosylation in cancer cell membranes as focused on glycolipids: overview and perspectives. Cancer Res. 45: 2405-2414.

34. Hakomori, S.I. (1988) In: Altered glycosilation in tumor cells. Editors: Ch. L. Reading, S.I. Hakomori, D.M. Marcus. New York: Alan R. Liss, p. 207-212.

35. Hakomori, S.I. (1989) Adv. Cancer Res. 52: 257-331.

36. Fenderson, B.A., Andrews, P.W., Nudelman, E., Clausen, H., Hakomori, S.I. (1987) Glycolipid core structure switching from globo to lacto-and ganglioseries during retinoic acid-induced differentiation of TERA-2-derived human embryonal carcinoma cells. Dev. Biol. 122: 21-34.

37. Fenderson, B.A., Eddy, E.M., Hakomori, S.I. (1990) Glycoconjugate expression during embryogenesis and its biological significance. BioEssays 12:173-79.

38. Lindenberg, S., Sundberg, K., Kimber, S.J., Lundblad, A. (1988) The milk oligosaccharide, lacto-N-fucopentaose l, inhibits attachment of mouse blastocysts on endometrial monolayers. J. Reprod. Fert. 83: 149-158.

39. Springer, G.F., Cheinsong-Popov, R., Schirrmacher, V., Desoi, P.R., Tegtmeyer, H. (1983) Proposed molecular basis of murine tumor cell-hepatocyte interaction. J. Biol. Chem. 258: 5702-5706.

40. Abe, K., Hakomori, S.I., Ohshiba, S. (1986) Differential expression of difucosyl type 2 chain (LeY) defined by monoclonal antibody AH6 in different locations of colonic epithelia, various histological types of colonic polyps, and adenocarcinomas. Cancer Res. 46: 2639-2644.

41. Itzkowitz, S.H., Shi, Z.R., Kim, Y.S. (1986) Heterogeneous expression of two oncodevelopmental antigens, CEA and SSEA-1, in colorectal cancer. Histochem. J. 18:155-163.

42. Itzkowitz, S.H., Yuan, M., Gukushi, Y., Palekar, A., Phelps, P.C., Shamsuddin, A.M., Trump, B.F., Hakomori, S.I., Kim, Y.S. (1986) Lewisx- and sialylated Lewisx-related antigen expression in human malignant and nonmalignant colonic tissues. Cancer Res. 46: 2627-2632.

43. Itzkowitz, S.H., Yuan, M., Ferrell, L.D., Palekar, A., Kim, Y.S. (1986) Cancer-associated alterations of blood group antigen expression in human colorectal polyps. Cancer Res. 46: 5976-5984.

44. Itzkowitz, S.H., Yuan, M., Montgomery, C.K., Kjeldsen, T., Takahashi, H.K., Bigbee, W.L., Kim, Y.S. (1989) Expression of Tn, sialosyl-Tn, and T antigens in human colon cancer. Cancer Res. 49: 197-204.

45. Kim, Y.S., Yuan, M., Itzkowitz, S.H., Sun, Q., Kaizu, T., Palekar, A., Trump, B.F., Hakomori, S.I. (1986) Expression of LeY and extended LeY blood group-related antigens in human malignant, premalignant, and nonmalignant colonic tissues. Cancer Res. 46: 5985-5992.

46. Clausen, H., Hakomori, S.I. (1989) Vox Sang. 56: 1-20.

47. Lloyd, K.O. (1988) Blood group antigen expression in epithelial tumors: influence of secretor status., In: Altered glycosylation in tumor cells. Editors: Ch.L. Reading, S.-I. Hakomori; D.M. Markus, NY. Alan R. Liss, pp. 235-243.

48. Stein, R., Chen, S., Grossman, W., Goldenberg, D.M. (1989) Human lung carcinoma monoclonal antibody specific for the Thomsen-Friedenreich Antigen. Cancer Res. 49:32-37.

49. Glinsky, G.V., Nikolaev, V.G., Ivanova, A.B., Shemchuk, A.S., Lisetsky, V.A. (1980) Experimental Oncology., V.2, N 1, p. 68-71.

50. Linetsky, M.D., Semyonova-Kobzar, R.A., Glinsky, G.V. (1991) The modifying influence of endogenous biopolymers, glycoamines from the blood serum of BALB/C mice on aggregation properties of cells of experimental rhabdomyosarcoma. Experimental Oncology 13, 6:27-33.

51. Hakomori, S.-I. (1991) Possible functions of tumor-associated carbohydrate antigens. Current Opinion in Immunology 3:646-653.

52. Hakomori, S.-I. (1992) Possible new directions in cancer therapy based on aberrant expression of glycosphingolipids: Antiadhesion and ortho-signalling therapy., Cancer Cells (in press)

53. Fung, P.Y.S., Madej, M., Koganty, R.R., Longenecker, B.M. (1990) Active specific innumotherapy of a murine mammary adenocarcinoma using a synthetic tumor-associated glycoconjugate. Cancer Res. 50:4308-4314.

54. Singhal, A., Fohn, M., Hakomori, S.I. (1991) Induction of a-N-Acetyllgalactosamine-O-Serine/Threonine (Tn) antigen-mediated cellular immune response for active immunotherapy in mice. Cancer Res. 51: 1406-1411.

55. Livingston, P.O., Natoli, E.J., Calves, M.J., Stockert, E., Oettgen, H.F., Old, L.J. (1987) Vaccines containing purified GM2 ganglioside elicit GM2 antibodies in melanoma patients. Proc. Natl. Acad. Sci. USA 84:2911-2915.

56. Livingston, P.O., Ritter, G., Srivastava, P. Padavan, M., Calves, M.J., Oettgen, H.F., Old, L.J. (1989) Characterization of IgG and IgM antibodies induced in melanoma patients by immunization with purified G_{M2} ganglioside. Cancer Res. 49:7045-7050.

57. Toyokuni, T., Dean, B., Hakomori, S.I. (1990) Synthetic vaccines: I. Synthesis of multivalent Tn antigen cluster-lysyllysine conjugates. Tetrahedron Lett. 31:2673-2676.

58. Ponpipom, M.M., Bugianesi, R.L., Robbins, J.C., Doebber, T.W., Schen, T.Y. (1981) Cell-specific ligands for selective drug delivery to tissues and organs. J. Med. Chem. 24:1388-95.

59. Fenderson, B.A., Zehavi, U., Hakomori, S. (1984) A multivalent lacto-N-fucopentaose III-lysyllysine conjugate decompacts preimplantation mouse embryos, while the free oligosaccharide is ineffective. J. Exp. Med. 160:1591-96.

60. Oguchi, H., Toyokuni, T., Dean, B., Ito, H., Otsuji, E., Jones, V.L., Sadoizai, K.K., Hakomori, S. (1990) Effect of lactose derivatives on metastatic potential of B16 melanoms cells. Cancer Communication 2:311-316.

61. Humphries, M.J., Olden, K., Yamada, K.M. (1986) A synthetic peptide from fibronectin inhibits experimental metastasis of murine melanoma cells. Science 233:467-470.

62. Iwamoto, Y., Robey, F.A., Graf, J., Sasaki, M., Kleinman, H.K., Yamada, Y., Martin, G.R. (1987) YIGSR, a synthetic laminin pentapeptide, inhibits experimental metastasis formation. Science 238:1132-1134.

63. Saiki, I., Iida, J., Murata, J., Ogawa, R., Nishi N., Sugimura, K., Tokura, S., Azuma, I. (1989) Inhibition of the metastasis of murine malignant melanoma by synthetic polymeric peptides containing core sequences of cell-adhesion molecules. Cancer Res. 49:3815-3822.

64. Watanabe, M., Ohishi, T., Kuzuoka, M., Nudelman, E.D., Stroud, M.R., Kubota, T., Kodaira, S., Abe, O.,

Hirohashi, S., Shimosato, Y., Hakomori, S. (1991) In vitro and in vivo antitumor effects of murine mono-clonal antibody NCC-SST-421 reacting with dimeric LeA (Leb/LeA) epitope. Cancer Res. 51:2199-2204.

65. Drickamer, K. (1991) Clearing up glycoprotein hor-mones [comment]. Clearing up glycoptotein hor-mones. Cell 67:1029-1032.

66. Kornfeld, R., Kornfeld, S. (1976) Comparative as-pects of glycoprotein structure. Annu. Rev. Biochem. 45: 217-37.

67. Sttahl, P.D. Rodman, J.S., Miller, M.J., Schlesinger, P.H. (1978) Evidence for receptor-mediated binding of glycoproteins, glycoconjugates and lysosomal gly-cosidases by alveolar macrophages. Proc. Nat. Acad. Sci. USA, 75: 1399-403.

68. Lee, Y.C. Carbohydr. Res. (1978) 67: 509-14.

69. Hormick, C.L., Kanish, F. (1972) Immunochemis-try 9:325-328.

70. Tam, J.P. (1988) Synthetic peptide vaccine design: synthesis and properties of a high-density multiple antigenic peptide system. Proc. Natl. Acad. Sci. USA, 85: 5409-13.

71. Tam, J.P., Lu, Y.A. (1989) Vaccine engineering: enhancement of immunogenicity of synthetic pep-tide vaccines related to hepatitis in chemically de-fined models consisting of T- and B-cell epitopes. Proc. Natl. Acad. Sci. USA, 86: 9084-8.

72. Posnett, D.N., McGrath, H., Tam, J.P. (1988) J. Biol. Chem. 263:1719-1725.

CHAPTER 10

CONCLUSION: PROSPECT FOR ANTI-CELL ADHESION THERAPY OF CANCER METASTASIS AND AIDS

The concept of "anti-adhesion" or "anti-aggregation" therapy of cancer, particularly cancer metastasis, has recently been proposed[1-6] with emphasis on inhibition of carbohydrate-mediated tumor cell aggregation and adhesion. As we have earlier summarized,[2] histogenetically different normal (embryonal, lymphoid) and a wide range of cancer cells and tissues express either transitorily (normal) or permanently (cancer), BGA-related glycodeterminants that are involved in the initial early stages of a multistep "cascade" cell adhesion mechanism. Like embryogenesis, tissue repair, regeneration and remodeling could be accompanied by expression of BGA-related glycoepitopes at definite stages of cell differentiation when "sorting-out" behavior of one homotypic cell population from a heterotypic assemblage of cells occurs. Different molecular isoforms of glycomacromolecules in serum contain cell-specific glycodeterminants which are involved in cell recognition, association and aggregation. These glycomacromolecules and serum carbohydrate-binding proteins, e.g., lectins, anticarbohydrate antibodies, etc., play opposite roles as inhibitors (low molecular weight glycoamines) and/or stimulators (high molecular weight multiglycoepitope-containing serum macromolecules and naturally occurring anticarbohydrate antibodies) of cell aggregation.

Initial observations have shown the effectiveness of these types of synthetic and natural molecules with biospecificity for glycodeterminants of cell adhesion as inhibitors of in vitro cell aggregation and potential antimetastatic agents in vivo.[1-25] It has been shown that specific carbohydrate units of glycosphingolipids, which are involved in cancer cell adhesion, as well as glycosphingolipid-containing liposomes, significantly suppressed B16 melanoma lung metastasis.[25] Effective inhibition of experimental metastasis has been shown by peptide cell adhesion determinants[26-28] as well as by monoclonal antibodies against lectins that participate in tumor cell aggregation (see above). However, the disadvantages of secondary immunological reaction and the nonspecific nature of these bioaactive molecules for tumor cells have blocked their potential therapeutic use.

Cell aggregation glycodeterminants has never been a target for antimetastatic drug development. But at least some of them have a unique design for anticancer drug potential: they have been detected in from 75% to more than 90% of human breast and lung carcinomas, and have never been detected in normal healthy tissue.[11,29] The use of monoclonal antibodies and/or active immunization against these glycodeterminants obviously has very limited potential, since, as we already have indicated,[2] serum of healthy individuals contain naturally occurring anticarbohydrate antibodies of the same specificity,[11,30] and immunoglobulins, particularly IgM, do not readily permeate blood vessels and have no access to extravascular tumor cells. It has been considered that naturally occurring anticarbohydrate antibodies may play a key role in cancer pathogenesis as a major immunoselective factor of metastasizing cancer cells during tumor progression,[2] and it has been shown that mAb against glycodeterminants of cancer cell aggregation did induce strong aggregation of human tumor cells.[31] The compounds which we propose to use as tools in anticancer therapy are essentially free from these limitations. Thus, this concept contains two important practical new elements: a) the tumor cell aggregation glycodeterminant as a target structure for antimetastatic drug development; and b) a new family of low molecular weight synthetic compounds with biospecificity for glycodeterminants of tumor cell aggregation as a tool for antimetastatic cancer therapy.

The one obvious limitation for this kind of cancer therapy has to be considered. Cancer is characterized by profound disturbances of cell adhesion, particularly of those stages that involve the protein adhesion systems (extracellular matrix proteins, integrins, immunoglobulin superfamily, gap junctional communications, etc). Aberrations in cancer cell adhesion are reversible, since the normal morphology of transformed cells in vitro can be restored by the addition of fibronectin[32-34] derived either from normal or transformed cells.[33] It seems that this fundamental property of cancer cells is critically important for invasion, dissemination of cancer cells from the primary tumor and their intravasation. Associated with neoplastic transformation, gross failure in junctional communication among the cancer cells is one of the key factors that probably cause uncontrolled cancerous growth.[36] It has been suggested that a decrease in homotypic adhesion between the transformed cells may support cancer cell release from the primary tumor and metastatic dissemination.[35] Furthermore, among hepatomas, low homotypic adhesion was found to be correlated with high metastatic potential.[37] Therefore, antiadhesion cancer therapy may not be applied when the primary tumor is present, since stimulation of dissemination of cancer cells from the primary tumor site may occur.

Human cancer is accompanied by changes in the quantitative and qualitative characteristics of serum macromolecules, generally in the direction of excessive accumulation of the activities that support tumor cell aggregation in blood vessels and metastasis development. (Figs. 1 & 2) These conditions specifically are changed by surgical elimination of the primary tumor, because as we have shown[22] in the early period after surgery (2-3 days), the level of glycoamines (the major cell aggregation and metastasis inhibitory component in serum) decreases dramatically in the serum of oncological patients. The changes in the levels of stimulatory macromolecules are supposedly less radical since some of them (anticarbohydrate antibodies) are present permanently in the serum of all healthy individuals,[11,30] and it is well-known that high molecular weight proteins in serum have a relatively long half-life. Application of cancer cell adhesion inhibitor(s) for treatment of cancer metastasis should be designed as an additional therapy following surgical elimination of the primary tumor with emphasis on prevention of homotypic and heterotypic cancer cell aggregation in the blood vessels, extravasation of cancer cells, formation of metastatic deposit(s) and secondary tumor foci in target organs.

L-selectin is constitutively functional, present at high levels on circulating, nonactivated, resting neutrophils, and mediates their attachment to cytokine-stimulated endothelium by presenting neutrophil carbohydrate ligands, e.g., sialyl Lewis X, to the vascular E- and P- selectins.[37] Similarly, the corresponding BGA-related glycodeterminants that are present constitutively on the surface of cancer cells may mediate their ad-

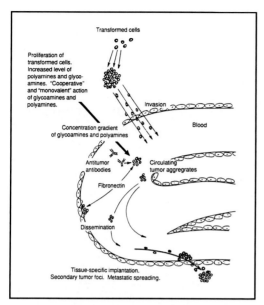

Fig. 1. Glycoamine-dependent mechanism of the dissemination of tumor cells.[1]

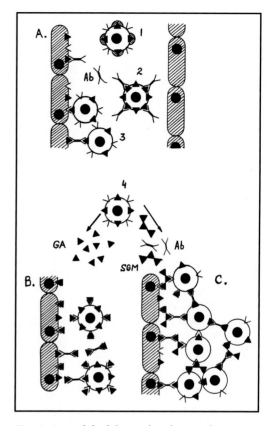

Fig. 2. A model of the molecular regulatory network controlling cancer cell aggregation and adhesion in the blood stream. Cancer cells which do not express cell adhesion glycodeterminants (1) will not be able to escape from the circulation into the tissue and would be killed shortly. Cancer cells with high density of cell adhesion glycodeterminants (2) would be targets for strong anticarbohydrate antibody attack and subsequent complement-dependent lysis. Cancer cells with a low (3) or medium (4) density of glycodeterminants of cell adhesion on the cell surface would be able to adhere to each other and/or to endothelium either directly (panel A) or through anticarbohydrate antibodies (Ab) bridges (panel B). The serum glycomacromolecules play an opposite role as inhibitors (monoglycoepitope-containing low molecular weight glycoamines-GA) or stimulators (multiglycoepitope-containing high molecular weight cancer serum markers and other macromolecules—SGM) of cancer cell aggregation and adhesion. Therefore, the progressive (panel C) or regressive (panel B) course of the metastatic disease may depend on the balance of corresponding stimulatory and inhibitory molecules in blood serum.

hesion to the endothelium and/or attachment at specific site(s) with leukocytes or platelets. However, in the absence of secondary activation-triggered stable attachment, or in the presence of adhesion inhibitor(s), this primary adhesion is reversible and arrested circulating cells will be released and returned to circulation.[38] Even activation-dependent adhesion via leukocyte integrins, which is stable for minutes under physiologic shear forces, is in principle reversible: neutrophils bound to endothelium by integrin-mediated adhesion (Mac-1) are released spontaneously after 10-15 minutes.[39] Thus, extravasation is not obligatory following even specific stable attachment of leukocytes to endothelium, and this may allow the time necessary for anti-cell aggregation action of a prospective metastasis inhibitor. The number of tumor cells in the lungs declined very rapidly after IV injection: after 24h 90-99% had disappeared[41-46] and after three days generally less than 1% remained.[42,45,46]

This decline is probably not due to a major redistribution of tumor cells, rather the cancer cells appear to degenerate rapidly.[42,47] Death of tumor is not due to their contact with immune host cells. Possibly they get damaged by the forces exerted by the blood stream[47] or other factors may cause a rapid degeneration of cancer cells in blood. Thus, the exceedingly low survival of tumor cells in the circulation is one of the the most important aspect of the metastatic process.[35,47] For example, a small difference in cell death (less than 1%) can lead to a large difference in metastasis formation (10 times or more).[35] Therefore, inhibition of extravasation of cancer cells, blocking of their homotypic and heterotypic adhesion would prevent escape of metastatic cells from blood into the tissues and may cause a dramatic reduction or even complete prevention of metastases development.

One of the intriguing conclusions from the concept presented here is that prospective leukocyte-endothelial cell adhesion inhibitor(s) which could be designed as an antiinflammatory agent may well be an effective antimetastatic and anti-AIDS drug. The simple first step for experimental verification of the above concept is also in the hands of vascular cell biologists: the monoclonal antibody that inhibits endothelial cell-leukocyte recognition and adhesion should also block in vivo cancer metastasis, and cancer cells may preferentially accumulate in vivo at the site of inflammation. The enhancement of distant metastases at the site of inflammation or injury[48-50] or at the site of surgical incisions, even those far from the place of tumor resection[51] is in agreement with the suggested conductor-like role of specific leukocyte subsets in metastasis since leukocytes accumulate selectively at the site(s) of injury or inflammation.

Obviously, the concept provides a theoretical basis for application of anticarbohydrate vaccine therapy for treatment of cancer and AIDS. The efficiency of an experimental cancer anticarbohydrate vaccine therapy has been reported[15,16] and the possibility of its application for AIDS treatment has been suggested.[52] However, the strategy of anticarbohydrate vaccine therapy that can be developed on the basis of our concept is quite different. In the case of cancer, it should be antimetastatic vaccine therapy only after surgical removal of the primary tumor. In the case of AIDS, it should be postexposure anticarbohydrate vaccine therapy. In both cases, this kind of therapy is supposed to be immunosuppressive because it will affect the immune cells trafficking, recognition, association and communication. However, we believe that in both cases (cancer and AIDS) such therapy will be appropriate since patients will need it briefly, and it will prevent the generalization of the disease. In the case of AIDS, this treatment will block the spread of HIV infection from primarily infected cells throughout the body, e.g., brain, lung, liver, etc. In the case of cancer, it will prevent metastasis by blocking of tumor cell aggregation in the blood stream, and it will affect immunoselection of the immunoresistant and highly metastatic clones of malignant cells during tumor progression.[2] The paradoxical efficiency of immunosuppressive therapy for treatment of experimental immunodeficiency syndrome has been shown recently.[53] To us, it seems more promising not to use antibodies and/or lectins as a tools of anticarbohydrate vaccine therapy, but glycoepitopes themselves linked to the appropriate low molecular weight carrier to increase bioaffinity.[2] The effectiveness of these types of synthetic and natural molecules as inhibitors of cell aggregation in vitro and in vivo has been recently suggested and demonstrated.[1-25]

REFERENCES

1. Glinsky, G.V. (1992) Glycoamines: Structural-Functional characterization of a new class of human tumor markers., In: Serological Cancer Markers. Editor: S. Sell. The Humana Press., Totowa, NJ, Chapter 11, p. 233-260.

2. Glinsky, G.V. (1992) The blood group antigens (BGA)-related glycoepitopes. A key structural determinants in immunogenesis and cancer pathogenesis. Critical Reviews in Oncology/Hematology 12:151-166.

3. Hakomori, S.-I. (1991) Possible functions of tumor-associated carbohydrate antigens. Current Opinion in Immunology 3:646-653.

4. Hakomori, S.-I. (1992) Possible new directions in cancer therapy based on aberrant expression of glycosphingolipids: Antiadhesion and ortho-signalling therapy. Cancer Cells (in press)

5. Glinsky, G.V. (1992) The site specificity of cancer metastasis: Is it determined by leukocyte-endothelial cell recognition and adhesion? (in preparation)

6. Glinsky, G.V. (1992) Glycodeterminants of melanoma cell adhesion. A model for antimetastatic drug design. Critical Reviews in Oncology/Hematology

(submitted)

7. Glinsky, G.V. (1990) Immunoselective hypothesis of tumor progression. Role aberrant glycosylation, anticarbohydrate antibodies, extracellular glycomacromolecules and glycoamines. J. Tumor Marker Oncology. 5:206.

8. Glinsky, G.V. (1990) Glycoamines, aberrant glycosylation and cancer: a new approach to the understanding of molecular mechanism of malignancy., In: Molecular Oncology. Oncodevelopment proteins and clinical applications. XVIIIth meeting of the International Society for Oncodevelopmental Biology and Medicine. Abstract Book, Moscow, USSR, September 23-27, 1990, p.7.

9. Glinsky, G.V., Semyonova-Kobzar, R.A., Berezhnaya, N.M. (1990) Modification of cellular adhesion, metastasizing and immune response by glycoamines: implication in the pathogenetical role and potential therapeutic application in tumoral disease. J. Tumor Marker Oncology. 5:231.

10. Fenderson, B.A., Andrews, P.W., Nudelman, E., Clausen, H., Hakomori, S.I. (1987) Glycolipid core structure switching from globo to lacto-and ganglioseries during retinoic acid-induced differentiation of TERA-2-derived human embryonal carcinoma cells. Dev. Biol. 122: 21-34.

11. Springer, G.F. (1984) T and Tn, general carcinoma autoantigens. Science 224: 1198-1206.

12. Springer, G.F., Cheinsong-Popov, R., Schirrmacher, V., Desoi, P.R., Tegtmeyer, H. (1983) Proposed molecular basis of murine tumor cell-hepatocyte interaction. J. Biol. Chem. 258: 5702-5706.

13. Fenderson, B.A., Eddy, E.M., Hakomori, S.I. (1990) Glycoconjugate expression during embryogenesis and its biological significance. BioEssays 12: 173-79.

14. Lindenberg, S., Sundberg, K., Kimber, S.J., Lundblad, A. (1988) The milk oligosaccharide, lacto-N-fucopentaose l, inhibits attachment of mouse blastocysts on endometrial monolayers. J. Reprod. Fert. 83, 149-158.

15. Fung, P.Y.S., Madej, M., Koganty, R.R., Longenecker, B.M. (1990) Active specific immunotherapy of a murine mammary adenocarcinoma using a synthetic tumor-associated glycoconjugate. Cancer Res. 50:4308-4314.

16. Singhal, A., Fohn, M., Hakomori, S.I. (1991) Induction of α-N-Acetyllgalactosamine-O-Serine/Threonine(Tn)antigen-mediated cellular immune response for active immunotherapy in mice. Cancer Res. 51:1406-1411.

17. Livingston, P.O., Natoli, E.J., Calves, M.J., Stockert, E., Oettgen, H.F., Old, L.J. (1987) Vaccines containing purified GM2 ganglioside elicit GM2 antibodies in melanoma patients. Proc. Natl. Acad. Sci. USA 84:2911-2915.

18. Livingston, P.O., Ritter, G., Srivastava, P. Padavan, M., Calves, M.J., Oettgen, H.F., Old, L.J. (1989) Characterization of IgG and IgM antibodies induced in melanoma patients by immunization with purified G_{M2} ganglioside. Cancer Res. 49:7045-7050.

19. Toyokuni, T., Dean, B., Hakomori, S.I. (1990) Synthetic vaccines: I. Synthesis of multivalent Tn antigen cluster-lysyllysine conjugates. Tetrahedron Lett. 31:2673-2676.

20. Glinsky, G.V., Linetsky, M.D., Ivanova, A.B., Osadchaya, L.P., Semyonova-Kobzar, R.A., Surmilo, N.I. (1990) Chemical, biochemical, spectroscopic and biological indentification of structural-functional determinants of glycoamines (aminoglycoconjugates) J. Tumor Marker Oncology 5:250.

21. Glinsky, G.V., Ogordniychuk, A.S., Shilin, V.V., Linetsky, M.D., Liventsov, V.V. Sidorenko, M.V., Surmilo, N.I., Kuchar, V.P (1990) Structural analysis of synthetic structural analogs of glycoamines: contribution into elucidation of the structure and biogenesis of natural aminoglycoconjugates. J. Tumor Marker Oncology 5:250.

22. Glinsky, G.V. (1989) Glycoamines: Biochemistry of a new class of humoral tumor markers. J. Tumor Marker Oncology 4:193-221.

23. Ponpipom, M.M., Bugianesi, R.L., Robbins, J.C., et al. (1981) Cell-specific ligands for selective drug delivery to tissues and organs. J. Med. Chem. 24:1388-95.

24. Fenderson, B.A., Zehavi, U., Hakomori, S. (1984) A multivalent lacto-N-fucopentaose III-lysyllysine conjugate decompacts preimplantation mouse embryos, while the free oligosaccharide is ineffective. J. Exp. Med. 160:1591-96.

25. Oguchi, H., Toyokuni, T., Dean, B., Ito, H., Otsuji, E., Jones, V.L., Sadoizai, K.K., Hakomori, S. (1990) Effect of lactose derivatives on metastatic potential of B16 melanoms cells. Cancer Communication 2:311-316.

26. Humphries, M.J., Olden, K., Yamada, K.M. (1986) A synthetic peptide from fibronectin inhibits experimental metastasis of murine melanoma cells. Science 233:467-470.

27. Iwamoto, Y., Robey, F.A., Graf, J., Sasaki, M., Kleinman, H.K., Yamada, Y., Martin, G.R. (1987) YIGSR, a synthetic laminin pentapeptide, inhibits experimental metastasis formation. Science 238:1132-1134.

28. Saiki, I., Iida, J., Murata, J., Ogawa, R., Nishi, N., Sugimura, K., Tokura, S., Azuma, I. (1989) Inhibition of the metastasis of murine malignant melanoma by synthetic polymeric peptides containing core sequences of cell-adhesion molecules. Cancer Res. 49:3815-3822.

29. Stein, R., Chen, S., Grossman, W., Goldenberg, D.M. (1989) Human lung carcinoma monoclonal antibody specific for the Thomsen-Friedenreich Antigen. Cancer Res. 49:32-37.

30. Lloyd, K.O., Old, L.J. (1989) Cancer Res., 49, 3445-3451.

31. Watanabe, M., Ohishi, T., Kuzuoka, M., Nudelman, E.D., Stroud, M.R., Kubota, T., Kodaira, S., Abe, O., Hirohashi, S., Shimosato, Y., Hakomori, S. (1991) In vitro and in vivo antitumor effects of murine monoclonal antibody NCC-SST-421 reacting with dimeric LeA (Leb/LeA) epitope. Cancer Res. 51:2199-2204.

32. Yamada, K.M., Yamada, S.S., Pastan, I. (1976) Cell surface protein partially restores morphology, adhesiveness, and contact inhibition of movement to transformed fibroblasts. Proc. Natl. Acad. Sci. USA, 73:1217-1221.

33. Hayman, E.G., Engwall, E., Ruoslahti, E. (1981) Concomitant loss of cell surface fibronectin and laminin from transformed rat kidney cells. J. Cell Biol. 88:352-357.

34. Ali, I.U., Mautner, V.M., Lanza, R.P., Hynes, R.O. (1977) Cell 11:115-126.

35. Ross, E. (1983) Cellular adhesion, invasion and metastasis. Biochim. Biophys. Acta 738:263-284.

36. Loewenstein, W.R. (1979) Junctional intercellular communication and the control of growth. Biochim. Biophys. Acta 560:1-65.

37. Hausman, R.E. (1983) Increase in homotypic aggregation of metastatic Morris hepatoma cells after fusion with membranes from nonmetastatic cells. Int. J. Cancer 32:603-608.

38. Picker, L.J., Warnock, R.A., Burns, A.R., Doerschuk, C.M., Berg, E.L., Butcher, E.C. (1991) The neutrophil selectin LECAM-1 presents carbohydrate ligands to the vascular selectins ELAM-1 and GMP-140. Cell 66:921-933.

39. Butcher, E.C. (1991) Leukocyte-endothelial cell recognition: three (or more) steps to specificity and diversity. Cell 67:1033-1036.

40. Lo, S.K., Detmers, P.A., Levin, S.M., Wright, S.D. (1989) Transient adhesion of neutrophils to endothelium. J. Exp. Med. 169:1779-1793.

41. Hewitt, H.B., Blake, A. (1975) Quantitative studies of translymphnodal passage of tumour cells naturally disseminated from a non immunogenic murine squamous carcinoma. Br. J. Cancer 31:25-35.

42. Fidler, I.J. (1970) Metastasis: guantitative analysis of distribution and fate of tumor embolilabeled with 125 I-5-iodo-2'-deoxyuridine. J. Natl. Cancer Inst.

45:773-782.

43. Proctor, J.W., Auclair, B.G., Rudenstam, C.M. (1976) The distribution and fate of blood-borne 125IUdR-labelled tumour cells in immune syngeneic rats. Int. J. Cancer 18:255-262.

44. Proctor, J.W. (1976) Rat sarcoma model supports both soil seed and mechanical theories of metastatic spread. Br. J. Cancer 34:651-654.

45. Weston, B.J., Carter, R.L., Easty, G.C., Connell, D.I., Davies, A.J.C. (1974) The growth and metastasis of an allografted lymphoma in normal, deprived and reconstituted mice. Int. J. Cancer 14:176-185.

46. Kodama, M., Kodama, T. (1975) Enhancing effect of hydrocortisone on hematogenous metastasis of Ehrlich ascites tumor in mice. Cancer Res. 35:1015-1021.

47. Roos, E., Dingemans, K.P. (1979) Mechanisms of metastasis. Mechanisms of metastasis. Biochim. Biophys. Acta, 560:135-166.

48. Sugarbaker, E.V., Cohen, A.M., Ketcham, A.S. (1971) Ann. Surg., 174:161-166.

49. Smith, R.R., Thomas, L.B., Hilbers, A.W. (1958) Do metastases metastasize? Cancer, 11:53-62.

50. Fisher, B., Fisher, E.R., Feduska, N. (1967) Trauma and the localization of tumor cells. Cancer, 20:23-30.

51. Der Hagopian R.P., Sugarbaker, E.V., Ketcham, A. (1978) Inflammatory oncotaxis. J. Am. Med. Assoc. 240:374-375.

52. Hansen, J.E.S., Clausen, H., Nielsen, C., Teglbjaerg, L.S., Hansen, L.L., Nielsen, C.M., Dabelsteen, E., Mathiesen, L., Hakomori, S.I., Nielsen, J.O. (1990) Inhibition of human immunodeficiency virus (HIV) infection in vitro by anticarbohydrate monoclonal antibodies: peripheral glycosylation of HIV envelope glycoptorein gp120 may be a target for virus neutralization. J. Virol. 64:2833-2840.

53. Simard, C., Jolicoeur, P. (1991) The effect of antineoplastic drugs on murine acquired immunodeficiency syndrome. Science 251:305-308.

INDEX